THE NEW PARADIGM IN SCIENCE

AND SYSTEMS THEORY

40 BESTSELLING BOOKS AND DVDS REVIEWED BY PIERRE F. WALTER

ILLUSTRATED EDITION

Published by Sirius-C Media Galaxy LLC

http://sirius-c-publishing.com

http://siriuscmedia.com

http://ipublica.com

ISBN 978-1-468149-48-7

Contact Information Pierre F. Walter

publisher@sirius-c-publishing.com

About Pierre F. Walter

http://drpfw.info

Quotation Suggestion

Pierre F. Walter, *The New Paradigm in Science and Systems Theory: 40 Bestselling Books and DVDs Reviewed,* Newark: Sirius-C Media Galaxy LLC, 2011

About the Author

Pierre F. Walter is an international lawyer, researcher, author and lecturer. After finalizing his studies in German law and European integration with diplomas in both disciplines in 1982, he graduated in December 1987 at the law faculty of the University of Geneva as *Docteur en Droit* in international law. The doctorate was funded by scholarships from the *Swiss Institute of Comparative Law*, Lausanne, and from the *University of Geneva*, as well as a Fulbright Travel Grant for an assistantship with Professor Louis B. Sohn at UGA Law School Department of International Law, Athens, Georgia, USA, in 1985. Pierre F. Walter also served as a research assistant to Freshfields, Bruckhaus, Deringer, Cologne, Germany in 1983 and to Lalive Lawyers, Geneva, in 1987.

Pierre F. Walter writes, lectures and teaches in English, German and French languages; he has written *more than ten thousand pages* embracing all literary genres, including *novels, short stories, film scripts, essays, selfhelp books, monographs* and extended *book reviews*. Also a pianist and composer, he has realized 40 CDs with jazz, newage and relaxation music.

Pierre F. Walter's professional publications span the domains *International Law, Criminal Law, Holistic Science, Psychology, Education, Shamanism, Ecology, Spirituality, Quantum Physics, Systems Theory, Natural Healing, Peace Research, Personal Growth, Selfhelp* and *Consciousness Research*.

110 Book Reviews, thirty-eight audio books and more than hundred video lectures were realized in the years 2005-2010. Besides, Pierre F. Walter is editor of a series 'Great Minds', which features scientists, artists and authors of genius from Leonardo to Fritjof Capra. In his 2011 series of scholarly articles, the author treats various topics from the realms of social science, psychology, mythology, medicine, oriental wisdom and psychoanalysis. In addition, in 2011, Pierre F. Walter published specially targeted readers, book reviews, on the New Paradigm in Business, Marketing and Career, and in Science and Systems Theory, as well as in Consciousness, Psychology, Healing and Spirituality

Pierre F. Walter publishes via his Delaware firm *Sirius-C Media Galaxy LLC* and the imprints IPUBLICA and Sirius-C Media (SCM).

To Fritjof Capra, who, if ever,
replied to me once in anger and fury
and put me off his agenda.

CONTENTS

Published by Sirius-C Media Galaxy LLC, 2011

Published by Sirius-C Media Galaxy LLC, 2011

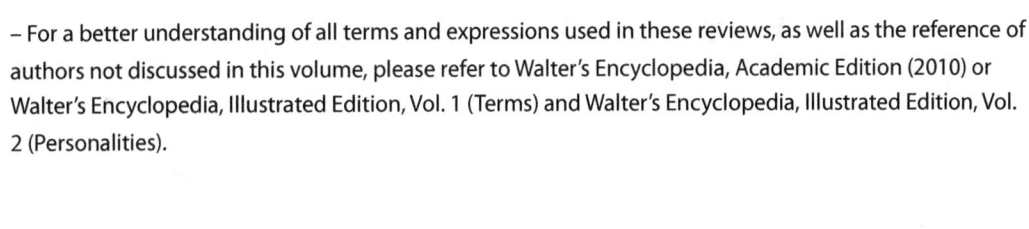

– For a better understanding of all terms and expressions used in these reviews, as well as the reference of authors not discussed in this volume, please refer to Walter's Encyclopedia, Academic Edition (2010) or Walter's Encyclopedia, Illustrated Edition, Vol. 1 (Terms) and Walter's Encyclopedia, Illustrated Edition, Vol. 2 (Personalities).

HAROLD SAXTON BURR

Harold Saxton Burr (1889-1973) was E. K. Hunt Professor Emeritus, Anatomy at Yale University School of Medicine. Burr was a member of the faculty of medicine for over forty-three years. From 1916 to the late 1950's, he published, either alone or with others, more than ninety-three scientific papers. Burr is most well known for his claim that all living things are molded and controlled by electro-dynamic fields, which could be measured and mapped with standard voltmeters. He named them fields of life or simply the *L-field*.

Beginning in the 1930s with H. S. Burr's seminal work at Yale, there has been a gradual accumulation of hard data to support the hypothesis of subtle energy fields that govern the human body. Burr set up a series of ingenious experiments, later repeated by other researchers, which demonstrated that all living organisms are surrounded and encompassed by their own energy fields, which he called *Life-fields*, and it's as a further shortening that this term later became *L-fields*.

He showed that changes in the electrical potential of the L-field would lead to changes in the health of the organism. By leaving some trees on the Yale campus hooked up to his L-field detectors for decades, he was able to show that changes in environmental electromagnetic fields, caused by such things as the phases of the moon, sunspot activity, and thunderstorms, substantially affected the L-field. He found he could detect a specific field of energy in a frog's egg, and that the nervous system would later develop precisely within that field, suggesting that the L-field was the organizing matrix for the body.

In his work with humans, he was able to chart and predict the ovulation cycles of women, to locate internal scar tissue, and to diagnose potential physical ailments, all through the reading of the individual's L-field. This latter finding, leading to his insistence that the L-field forms primary to the physical, would eventually have Burr vilified for *wishful vitalism*.

Student and colleague Leonard Ravitz carried Burr's work forward. Ravitz focused especially on the human dimension, beginning with a solid demonstration of the effects of the lunar cycle on the human L-field, reaching a peak of activity at the full moon. Through work with hypnotic subjects, he demonstrated that changes in the L-field directly relate to changes in a person's mental and emotional states.

> **Leonard Ravitz**
>
> Both emotional activity and stimuli of any sort involve mobilization of electrical energy, as indicated on the galvanometer, hence, both emotions and stimuli evoke the same energy. Emotions can be equated with energy.

Most intriguingly, Ravitz demonstrated that the L-field as a whole disappears before physical death. While Burr expressed himself in a rather misleading terminology, speaking of 'electricity' when he connoted that vital or life energy, and of 'electromagnetic fields' when it's all about *The Field* (Lynn Taggart), and thus the bioplasmatic life energy (that Masaru Emoto calls *hado*, which could be translated as 'vibration', these hassles about terminology could not prevent most of the literature on energy and vibrational medicine to cite Burr as one of their pioneers. Masaru Emoto points out in his book *The Secret Life of Water (2005)*, p. 139:

> **Masaru Emoto**
>
> About seven decades ago, a scientist named Harold Saxton Burr laid much of the basic foundation for the science of hado. Burr was a renowned professor of anatomy at Yale University. In his

attempt to understand the mysteries of life, he gave us the term *L-field* or *life field*. Since all the cells without our bodies are replaced over a period of six months, why do we keep being reborn as the same person over and over? Like a mold used to make Jell-O, an invisible force enables this to happen, he believed, and he called it the 'life field'. He believed that since the life field is an electrical field in nature, it could be measured, and he even developed his own measuring device using a voltage indicator and an electrode. He discovered that the measurements he took varied with the way the subject was feeling. He got higher voltage readings from subjects who were feeling blissful, and lower voltage readings from those feeling depressed.

Published by Sirius-C Media Galaxy LLC, 2011

HERBERT JAMES CAMPBELL

Herbert James Campbell, a renowned English neurologist, found in twenty-five years of research a universal principle which dominates our brain: *the pleasure principle*. This sounds like Freud, but it has little to do with psychoanalysis or psychology. What we are

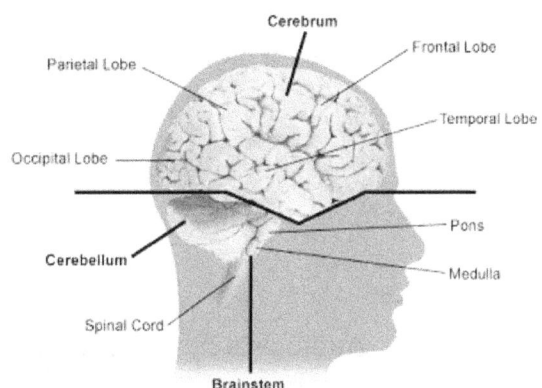

facing here are facts proven by natural science, by neurology. In 1973 Campbell published his book *The Pleasure Areas* which represents a summery of many years of neurological research.

Campbell succeeded in demonstrating that our entire thinking and living is primarily motivated by pleasure. Pleasure not only in a tactile-sensuous or sexual way, but also as non-sensuous, intellectual or spiritual pleasure. With these findings, the old theoretical controversy if man was primarily a biological or a spiritual being became obsolete. For it is in the first place *our striving for pleasure* that induces certain interests in us, that drives us to certain actions and that lets us choose certain ways.

During childhood and depending on the outside stimuli we are exposed to, certain *preferred pathways* are traced in our brain, which means that specific neural connections are established that serve the information flow. The number of those connections is namely an indicator for intelligence. The more of those preferred pathways exist in the brain of a person, the more lively appears that person, the more interested she will be in different things, and the quicker she will achieve integrating new knowledge into existing memory.

High memorization, Campbell found, is namely depending on how easily new information can be added-on to existing pathways of information. Logically, the more of those pathways exist, the better! Many preferred pathways make for high flexibility and the capacity to adapt easily to new circumstances. Campbell's research indicates that the repression of pleasure that is since centuries part of our Judeo-Christian culture, has negatively infringed upon human evolution and impaired the integrity of our psycho-somatic health.

Not only neurologists such as Campbell have nowadays thought about the basic functions of life and living, but also people who were formerly active in totally different fields of science. The American scientists Ashley Montagu and James W. Prescott had very different points of departure for their extensive research. Montagu wanted to know why small rhesus apes died when they were deprived from their mother while they survived when a simple felt mat was put in the cage as surrogate of motherly tactile affection. Prescott researched on the origins of violence. He did from the start not acknowledge the age-old pretension that man was per se a violent creature even though human history, or what historians saw of it, seemed to prove it.

Besides peace research, Campbell's findings are important for research on perception and the human memory surface. Our brain adds new information on to already existing information, most of the time, instead of forming a new pattern in the memory surface. This is how the brain, and the process of thought, works, and how this system impacts upon perception by actually per se distorting perception.

Some neurologists, and Campbell is among them, gave comprehensive answers. He argues that our brain has developed this kind of faulty memory surface because it was enhancing human evolution as a matter of survival -

Published by Sirius-C Media Galaxy LLC, 2011

while of course it has brought about millions of deficient thinkers! One of our major thinking trainers and international coaches, *Edward de Bono*, said the same from his own research on perception and his experience as a corporate trainer.

FRITJOF CAPRA

Books Reviewed

I found Capra's *Tao of Physics* in 1985, at a time when my life was in a complete re-orientation. In this situation, Capra's books *Tao of Physics* and *The Turning Point* reflected and emphasized the turning point in my own life. *Tao of Physics* hit me like a lightning, but what a blessing was that hit! The impact of the book on my psyche and my personal evolution was comparable only to the discovery of the *I Ching* and *Taoism*, as well as the writings and psychoanalytical teaching of *Françoise Dolto* that I equally discovered in 1985.

But today, having read all of Capra's publications, I think that his best book is *The Hidden Connections*. Reading Capra, I eventually found all my intuitions confirmed, and all the criticism that already as a child and school boy I had voiced against Western society and education, intelligently and comprehensively put together in a structural framework for the purpose of creating a *new reality*, a reality that will be holistic and emotionally as well as *erotically intelligent*. In fact, it was at that time that I devoted myself to serving humanity unconditionally to help creating this new reality, and to contribute to this mission voluntarily and with a transpersonal motivation. Capra has a unique gift of genius to formulate and explain complex scientific and philosophical insights and interrelations in a way that the educated reader can understand. Originally from Austria and brought up with German as his mother tongue, he learnt English so perfectly that from the moment he moved to Berkeley, California for his work as a quantum physicist, he wrote and published only in English. The parallels here are evident with *Albert Einstein* and *Wilhelm Reich* who equally were from Germanic origin and after their emigration to the United States only wrote and published in English. And from their level of genius and stringent originality, these three men can well be compared.

There are other important facts about Capra that are perhaps hardly known, and partly explain why he has this phenomenal lucidity, while he works as a mainstream scientist and yet in his books by far surpasses the limitations of this profession and the worldview of most of his professional colleagues (except those on his own level of genius).

Capra said somewhere in his books that he was raised in a quite matriarchal environment, an environment virtually deprived of males. He was raised by three women, and they were all single, for different reasons: his mother, his grandmother and his great grandmother. And they lived together with many animals on the big farm in the country. And Capra grew up in a probably lucky and happy childhood environment without having suffered abuse. All this is important, I think, in order to understand his basically non-judgmental worldview and his ability to understand people from ultra-orthodox to very liberal with the same generosity and magnanimity. Capra is truly exceptional in this respect. This can be seen in his lesser known volume *Uncommon Wisdom* which is a recollection of conversations with remarkable people, and at the same time a kaleido- scope of anecdotes form the life of a truly lively and communicative human being. The other noteworthy instance from Capra's life is his long involve- ment in the counter culture and his meeting with most of the celebrities of that culture, as for example Timothy Leary, Terence McKenna, Gregory Bate- son, Ronald David Laing or Thomas Szasz, the founders of the *Antipsychiatry* movement.

Besides Capra's intellectual brilliance and exqui- site use of language, it's the simplicity of his lit- eracy, and his unpretentious way to relate other people's achievements and remarkable traits with a certain modesty that let Capra stand out as a truly universal and encyclopedic scholar. The fact that his books have become worldwide bestsellers over many years, and were translated in all major languages of the world doesn't really surprise when you see that this scholar is way  beyond the proverbial dry scholar and enjoys a large platform of worldwide human connections and backup. In my view, it's Capra's extraordinary hu- man skills, his ability to communicate, and communicate across scientific disciplines together with a strongly integrative mindset and attitude that made him such an important alternative figure in the mainstream science environment.

Capra is one of the most important holistic thinkers of our times, and per- haps even *the* most important of our science philosophers today. His genius in no way goes second to Einstein's, while his language and appearance is

Published by Sirius-C Media Galaxy LLC, 2011

much more modest than that of many of our self-labeled new science gu-rus.

Fritjof Capra

The Tao of Physics

An Exploration of the Parallels between Modern Physics
and Eastern Mysticism
New York: Bantam Books, 1984

It is an almost impossible quest, if you refuse to be shallow, to write book reviews for each of Capra's books. That's because he expresses himself so well, and so completely that it sounds almost hubristic to attempt to paraphrase or report any of his ideas. You can't say it better than he said it, and that is a frustrating experience for a book reviewer. But after all, it's not impossible to try, and this is what I am going to do here!

I try to write a decent book review, and not quote too much of the text, because otherwise I will end up quoting pretty much the whole book. In fact, the quote collection I have drafted from my side remarks of the first five of Capra's books counts 144 pages, by itself a book.

So, how can I convey the thrill I experienced when reading Capra's *Tao of Physics?* How can I even remotely convey with my own words what Capra wrote about, as it's actually something you can hardly put in words. Of course, much has been written about this book, and it is always quoted as the ultimate synthesis between modern physics and ancient mysticism. But how did Capra come to prove his point? This is exactly the question. He did

Published by Sirius-C Media Galaxy LLC, 2011

it by proving a large number of small points, one after the other, and that is how he finally could prove the whole of his thesis or theory. This is his unique genius not only as a physicist and emotionally very intelligent man, but also as a writer. How would you set out to write about such vast a topic? I remember well to have discarded out writing a whole range of books that I had planned to write, because their topic was of similarly vast dimensions, and the terrain of research was similarly shaky as the terrain this book is based upon. Interestingly enough, the book did a miracle in that the terrain that was shaky at the very time the book was written, was no more shaky *after* the book was written. If this is not genius, I don't know *what* is genius. Capra has shifted a worldview by writing a book. I think that before him the only human who has accomplished this was Leonardo da Vinci, while he did it with painting and machines, not with writing, but this might be the reason why Capra's newest book is about nobody else than da Vinci.[1]

I hope that publishers and editors, and Capra himself, will not frown upon me for the abundant quotes in my reviews of his books. It's simply that I can't paraphrase what he so perfectly put in words, and I will restrain myself to a few comments that, hopefully so, are somewhat comprehensive. To begin with, Capra writes:

Fritjof Capra

If physics leads us today to a world view which is essentially mystical, it returns, in a way, to its beginning, 2500 years ago. It is interesting to follow the evolution of Western science along its spiral path, starting from the mystical philosophies of the early Greeks, rising and / unfolding in an impressive development of intellectual thought that increasingly turned away from its mystical origins to develop a world view which is in sharp contrast to that of the Far East. In its most recent stages, Western science is finally overcoming this view and coming back to those of the

---

1 See Fritjof Capra, *The Science of Leonardo (2008).*

early Greek and the Eastern philosophies. This time, however, it is not only based on intuition, but also on experiments of great precision and sophistication, and on a rigorous and consistent mathematical formalism./5-6

An important discourse in *The Tao of Physics* was Capra's report about the Eleatic school because it gives us an important clue for the origins of our intellectual dualism, a subject that Joseph Campbell has largely discussed in *Occidental Mythology*.

> **Fritjof Capra**
>
> The split of this unity began with the Eleatic school, which assumed a Divine Principle standing above all gods and men. This principle was first identified with the unity of the universe, but was later seen as an intelligent and personal God who stands above the world and directs it. Thus began a trend of thought which led, ultimately, to the separation of spirit and matter and to a dualism which became characteristic of Western philosophy./7

Capra's book is of course directed against that very dualism, and represents an attempt to overcome that schizoid split by showing that upon a deeper regard a *synthesis* between left-brain Western scientific thought and right-brain Eastern philosophy is the only intelligent way out of the dilemma caused by the stringent paradoxes of quantum mechanics. What I call in my writings the *schizoid split* in the internal setup of Western culture, Capra called it the division between spirit and matter:

> **Fritjof Capra**
>
> As the idea of a division between spirit and matter took hold, the philosophers turned their attention to the spiritual world, rather than the material, to the human / soul and the problems of ethics. These questions were to occupy Western thought for more

Published by Sirius-C Media Galaxy LLC, 2011

than two thousand years after the culmination of Greek science and culture in the fifth and fourth centuries B.C./6-7

Capra left no doubt that it was not the Eleatic School of thought that ultimately was going to cement that dualism in our cultural credo for the next two thousand years, backed up by the powerful rhetoric of Aristotle:

Fritjof Capra

The scientific knowledge of antiquity was systematized and organized by Aristotle who created the scheme which was to be the basis of the Western view of the universe for two thousand years. But Aristotle himself believed that questions concerning the human soul and the contemplation of God's perfection were much more valuable than investigations of the material world. The reason the Aristotelian model of the universe remained unchallenged for so long was precisely this lack of interest in the material world, and the strong hold of the Christian church which supported Aristotle's doctrines throughout the Middle Ages./8

And the next step, then, in the building of that cultural paranoia was the turn of events starting with the reductionist science philosophy of the French philosophers La Mettrie and René Descartes:

Fritjof Capra

The birth of modern science was preceded and accompanied by a development of philosophical thought which led to an extreme formulation of the spirit/matter dualism. This formulation appeared in the seventeenth century in the philosophy of René Descartes who based his view of nature on a fundamental division into two separate and independent realms: that of mind (res cogitans), and that of matter (res extensa). The Cartesian division allowed scientists to treat matter as dead and completely separate from themselves, and to see the material world as a multitude of different objects assembled into a huge machine./8

What Capra was showing here was the missing link between our modern-day separative and highly individualistic Western worldview, and its historical origins. And it explains very conclusively why we are torn up inside, fragmented and unwhole (unholy):

Fritjof Capra

This inner fragmentation mirrors our view of the world outside, which is seen as a multitude of separate objects and events. The natural environment is treated as if it consisted of separate parts to be exploited by different interest groups. The fragmented view is further extended to society, which is split into different nations, races, religions and political groups./8

After having shown how our fragmented Western worldview came about historically, Capra presents the Eastern worldview:

Fritjof Capra

In contrast to the mechanistic Western view, the Eastern view of the world is organic. For the Eastern mystic, all things and events perceived by the senses are interrelated, connected, and are but different aspects or manifestations of the same ultimate reality. (...) In the Eastern view, then, the division of nature into separate objects is not fundamental and any such objects have a fluid and ever-changing character. The Eastern / world view is therefore intrinsically dynamic and contains time and change as essential features. The cosmos is seen as one inseparable reality - forever in motion, alive, organic; spiritual and material at the same time./10-11

The danger of fragmentation, Capra explains very conclusively, is that we try to find absolute points of reference behind each of our fragmented concepts, and we do this probably unconsciously in an attempt to heal our inner split. Yet ultimately by doing this we bring about a distorted perception

Published by Sirius-C Media Galaxy LLC, 2011

of reality, by taking the proverbial finger that points to the moon, for the moon:

Fritjof Capra

For most of us it is very difficult to be constantly aware of the limitations and of the relativity of conceptual knowledge. Because our representation of reality is so much easier to grasp than reality itself, we tend to confuse the two and to take our concepts and symbols for reality. It is one of the main aims of Eastern mysticism to rid us of this confusion. Zen Buddhists say that a finger is needed to point to the moon, but that we should not trouble ourselves with the finger once the moon is recognized./15

In addition, facing the paradoxical behavior of electrons in the quantum world, Capra asked the intelligent question why Westerners are so terribly confused, even shocked, when encountering a paradox, or simply a completely illogical behavior? He found the answer in comparing Western thought with Eastern philosophy:

Fritjof Capra

Eastern mysticism has developed several different ways of dealing with the paradoxical aspects of reality. Whereas they are bypassed in Hinduism through the use of mythical language, Buddhism and Taoism tend to emphasize the paradoxes rather than conceal them./35

I think this difference between Indian thinking and the Chinese and Japanese philosophical traditions is very important, as Joseph Campbell has emphasized it as well in his book *Oriental Mythology (1962/1992)*.

The Zen tradition, derived from its original Chinese root philosophy (where it was called *Chan Buddhism*), is very fond of putting the stress on the paradox for a simple reason: the paradox teaches us the limitations of rational thinking and thereby shows us the *relativity* of a purely rational worldview. By seeing our obvious limitation, we can go beyond it and develop a more holistic, encompassing worldview, a worldview namely that gives the necessary space for the irrational, for the fantastic, the imaginative and the scurrilous in nature, and also in our human nature. Without the latter, humor, for example, as an expression of true humanity, is not possible. Obsessively rational thinkers are the greatest bores on earth – and besides, they are not very smart after all. They can't because the greater part of human wisdom simply escapes their squared minds.

This fundamental change in how we perceive reality as Western scientists is so important not for our better use of leisure time, but primarily because our whole science is going to shift, and must shift, according to this reorientation of the thinker. Capra makes it very clear that we cannot stay with the old Newtonian demons:

Fritjof Capra

The mechanistic view of nature ... is closely related to a rigorous determinism. The giant cosmic machine was seen as being completely causal and determinate. All that happened had a definite cause and gave rise to a definite effect, and the future of any part of the system could - in principle - be predicted with absolute certainty if its state at any time was known in all details. (...) The philosophical basis of this rigorous determinism was the fundamental division between the I and the world introduced by Descartes. As a consequence of this division, it was believed that the world could be described objectively, i.e., without ever mentioning the human observer, and such an objective description of nature became the ideal of all science./45

The result was that we discarded nature out of science and by doing this, created a fundamentally nature-hostile science, a science that destroys us by destroying our planet. This science, then, reflected exactly the distorted

Published by Sirius-C Media Galaxy LLC, 2011

view prevalent since Biblical times in Western culture, that the male is supe-rior to the female. This cult of male supremacy, as it were, led straight to a never-ending course of violence that slowly but definitely suffocates us. Capra writes:

> Fritjof Capra
> Western society has traditionally favored the male side rather than the female. Instead of recognizing that the personality of each man and of each woman is the result of an interplay be-tween male and female elements, it has established a static order where all men are supposed to be masculine and all women feminine, and it has given men the leading roles and most of society's privileges. This attitude has resulted in an over-emphasis of all the yang - or male - aspects of human nature: activity, rational thinking, competition, aggressiveness, and so on. The yin - or female - modes of consciousness, which can be described by words like intuitive, religious, mystical, occult, or psychic, have constantly been suppressed in our male-oriented society./133

> And the same biased perception of reality, distorting the har-mony between the male and the female principle, is to be seen throughout Western philosophy, in its abysmal dualism, which lacks the fundamental ability to find the synthesis that Oriental thought is so apt to establish. Capra agrees with the Eastern view that say all opposites are complementary and 'merely different aspects of the same phenomenon'. Capra wistfully remarks that in the East, 'a virtuous person is therefore not one who under-takes the impossible task of striving for the good and eliminating the bad, but rather one who is able to maintain a dynamic bal-ance between good and bad./131

When you look at the *Tao of Physics* from this perspective, from the big pic-ture behind the details of quantum physics, you will see that Capra's deeper message in this revolutionary book goes way beyond a redefinition of modern physics. Capra has prepared the ground in this earliest of his books for the giants to come, and he has given birth to the first giant, *The Turning Point*, only a couple of years after publishing *The Tao of Physics*.

While *The Tao* remains Capra's most popular book it is not his best book. The genius trick was that he developed the original idea further and found something like a new holistic concept for all sciences, but was never arrogant enough to label it fashionably as 'A Theory of Everything'.

Capra calls his new concept *ecological*, and while he has not invented that term, he surely has given a much broader content to it than it had ever before.

Published by Sirius-C Media Galaxy LLC, 2011

Fritjof Capra

The Turning Point

Science, Society and the Rising Culture
New York: Simon & Schuster (Flamingo), 1987

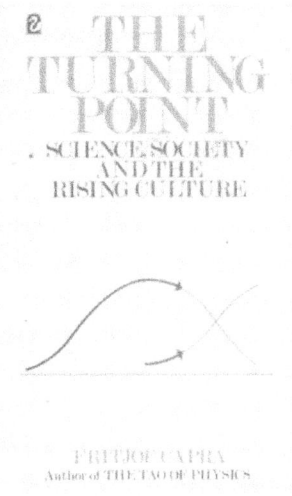

The Turning Point is one of Capra's most important books, and truly it was a turning point also in Capra's own life. In my personal view, and contrary to what most critics say, it is lesser the *Tao of Physics* that is the real strike of genius, but the present book. And this because of the extrapolation of the holistic concepts developed in the Tao upon the whole of international culture.

Only a thinker who is both logically precise, very knowledgeable about science history, and who has a metarational and integrated perception of the universe could do such a giant work. As a matter of coincidence, this book was marking also a turning point in my own life, and I found it, like a blessing, in a time of virulent contradictions and emotional turmoil in my life, back in 1985. And the cover that I have scanned for this review may give an idea how many times I have read this book and how intensely I commented it. In fact, the quotes I have taken from it go over thirty-two pages! This book is a must-read for a modern intellectual and I will simply not even talk or converse with anybody who pretends to be a progressive thinker and has *not* read *The Turning Point*.

The following quote shows the general direction that Capra took from this book, and that will be especially present in his two subsequent books, *The Web of Life*, and *Hidden Connections*. It has been called *the systems view*; it simply is a sound holistic science paradigm that can be practically applied to all scientific research, and that promises to bring about scientific, social and later political results that are in accordance with human dignity, fostering the expansion of human consciousness and evolution. These solutions namely will be different from those we had in the past because they will be integrated and sustainable, and this both in the fields of science and culture:

> **Fritjof Capra**
>
> These problems (…) are systemic problems, which means that they are closely interconnected and interdependent. They cannot be understood within the fragmented methodology characteristic to our academic disciplines and government agencies. Such an approach will never solve any of our difficulties but will merely shift them around in the complex web of social and ecological relations. A resolution can be found only if the structure of the web is changed, and this will involve profound transformations of our social institutions, values, and ideas./6

Capra carefully initiates the reader in the fact that his view is not just theory, but bears a direct significance for our daily life, and our daily problems:

> **Fritjof Capra**
>
> Studies of periods of cultural transformation in various societies have shown that these transformations are typically preceded by a variety of social indicators, many of them identical to the symptoms of our current crisis. They include a sense of alienation and an increase in mental illness, violent crime, and social disruption, as well as an increased interest in religious cultism – all of which have been observed in our society during the past decade. In times of historic cultural change these indicators have tended to appear one to three decades before the central transformation, rising in frequency and intensity as the transformation is approaching, and falling again after it has occurred./7

Published by Sirius-C Media Galaxy LLC, 2011

One of the points that make Capra a genius is his mental flexibility. Contrary to many other scientists from the so-called *exact* scientific disciplines, he has an extraordinarily synthetic thought capacity which makes him sense shifts and developments in society long before they actually happen. Then, following his intuition, he puts his sharp rational mind in the forefront so as to collect and arrange the information he needs to elucidate and deploy what he intuitively presents.

This is in accordance with Einstein's famous saying that a problem can never be solved on the level it was created. In fact, it's only through creative thinking and intuition that we can find new solutions to our old problems, because we then situate the thinker on a different level of perspective. This can be seen in the way Capra puts spotlights on trends and philosophical movements of long ago, to show the potential they have for shifting our view and preparing new ground for new, effective and unconventional solutions. Heraclites is one of the earliest geniuses who have shown us the way to go, but he was not followed. Instead Western science was to slavishly follow Aristotle, and in the East, the same happened when Lao-tzu was shunned by Chinese thinkers for giving the preference to the pedant, moralist and hair-splitting Confucius.

And truly so, presently, while we got all kinds of fancy developments in modern science in the aftermath of *What The Bleep Do We Know!? (2005/ 2006)*, and *Theories of Everything* virtually sprouting out of every university faculty, we are in midst of one of the most fatally rigid and most ignorant political developments ever to be seen in a democratic country. I can't think of any moment in history when the flourishing of science and the arts, material prosperity and abundance, and the intellectual prowess of a nation was in greater contradiction with the setup of its reigning political power structure, as it is the case in present-day America. And if we are unlucky, all of us, including our most gifted thinkers and scientists will be swept away by this earthquake of moronic fundamentalism and fascism that the present world leaders are busily preparing. And if we are lucky and there will be a real turning point, then Dr. Capra will surely become the first consultant of that future world government that we so badly need for the well-being of humanity.

I got to know about the Simontons and their extraordinarily successful alternative cancer therapy through Capra, through this present book, and would otherwise perhaps never have encountered them. And I have seen television interviews with Capra and could learn how this man communicates with other scientists and how he is capable, because of his modesty, to unite scientists from the most diverse departments into one common vision. It became clear from that documentary that Capra doesn't quote from a science publication without contacting the author and getting in a fruitful personal dialogue with them. The result is highly accurate, precise and trustworthy information in his books about all and every scientist, healer, philosopher or artist that he references.

One important area where the reigning paradigm is presently shifting is *psychology*. This is only now really apparent, in 2008, where we can count the books written about what today is called *Energy Psychology*, but at the time Capra authored *The Turning Point*, this was unthinkable. I have seen it myself when, after finalizing my *doctor of law* degree at the University of Geneva, in 1987, I was shifting majors and started to study psychology. And the first semester consisted of sixty percent statistics, and a few lectures about basic psychological terminology and ways of research. There was no word about psychoanalysis, no word about any holistic or systemic area of research, and it was boring me to death. I always thought that law is a dry subject to study but after having peeked in psychology, I can say that law is one of the most colorful and passionate subjects I have ever studied! Capra explains why the systems view of life will have a profound impact upon psychology:

Fritjof Capra

As in the new systems biology, the focus of psychology is now shifting from psychological structures to the underlying processes. The human psyche is seen as a dynamic system involving a variety of functions that systems theorists associate with the phenomenon of self-organization. Following Jung and Reich, many psychologists and psychotherapists have come to think of mental dynamics in terms of a flow of energy, and they also believe that these dynamics reflect an intrinsic intelligence - the equivalent of the systems concept of mentation - that enables the psyche not only to create mental illness but also to heal itself. Moreover, inner growth and self-actualization are seen as essential to the dynamics of the human psyche, in full agreement with the emphasis on self-transcendence in the systems view of life./407

In fact, one of Capra's friends is *Stanislav Grof*, and with Grof he has discussed many of the topics around psychology/psychiatry he writes about and that he has truly thought through until the end. I got this information not only from the huge footnote section in the present book, but also from his insightful book *Uncommon Wisdom*, in which he published interviews with leading edge personalities from all walks of life, and that stands as an example for Capra's extraordinary communication abilities.

Fritjof Capra
Uncommon Wisdom
Conversations with Remarkable People
New York, Bantam, 1989.

As I have to limit myself in this review to a few topics from the extraordinarily rich array of scientific disciplines Capra reviews in this book, I shall present, as an example, how he summarizes the alternative cancer healing approach developed by *O. Carl Simonton* and his wife, Stephanie Matthews-Simonton. Interestingly so, I have had this information thus about twenty years earlier, but it was only recently that I ordered their book *Getting Well Again (1978)*, and after having read it, I must admit that there was nothing essentially new I was learning through reading it. All the cutting-edge information about the book and the approach itself had been given by Capra:

Fritjof Capra

The popular image of cancer has been conditioned by the fragmented world view of our culture, the reductionist approach of our science, and technology-oriented practice of medicine. Cancer is seen as a strong and powerful invader / that strikes the body from outside. There seems to be no hope of controlling it, and for most people cancer is synonymous with death. Medical treatment - whether radiation, chemotherapy, surgery, or a combination of these - is drastic, negative, and further injures the body. Physicians are increasingly coming to see cancer as a systemic disorder; a disease that has a localized appearance but has the ability to spread, and that really involves the entire body, the original tumor being merely the tip of the iceberg./388-389

What many physicians and mainstream cancer researchers hide or veil is the fact that the strangeness of the current cancer therapy approach has nothing specific about it, and can be well explained, and criticized, by seeing through its ineffectiveness, its *Cartesianism*, its mechanistic and inhuman approach to healing, which is not healing in fact, but *medical business*. And it's a worldwide and gigantic business, and all the huge profits go in the hand of a few multinationals that use thousands of ignorant or uncritical medical doctors as their brave and brainwashed business consultants. Needless to add that it's one of the most sordid businesses in the world as it brings about huge misery, and thousands dying every year as a result of the

Published by Sirius-C Media Galaxy LLC, 2011

hypnotic spells of doctor-executioners that play the devil's agent, people who, what now is clear after twenty years of alternative cancer research, would not have needed to die in the first place! This is truly a shame considering the effectiveness of our perennial natural healing techniques. Capra writes in more hopeful terms, when he reports the Simonton approach to cancer healing. But the fact alone that the Simontons are successful in their approach shows with the best possible evidence that they must be right somehow:

Fritjof Capra

One of the main aims of the Simonton approach is to reverse the popular image of cancer, which does not correspond to the findings of current research. Modern cellular biology has shown that cancer cells are not strong and powerful but, on the contrary, weak and confused. They do not invade, attack, or destroy, but simply overproduce. A cancer begins with a cell that contains incorrect genetic information because it has been damaged by harmful substances or other environmental influences, or simply because the organism will occasionally produce an imperfect cell. The faulty information will prevent the cell from functioning normally, and if this cell reproduces others with the same incorrect genetic makeup, the result will be a tumor composed of a mass of these imperfect cells. Whereas normal cells communicate effectively with their environment to determine their optimal size and rate of reproduction, the communication and self-organization of malignant cells are impaired. As a result they grow larger than healthy cells and reproduce recklessly. Moreover, the normal cohesion between cells may weaken and malignant cells may / break loose from the original mass and travel to other parts of the body to form new tumors - which is known as metastasis. In a healthy organism the immune system will recognize abnormal cells and destroy them, or at least wall them off so they cannot spread. But if for some reason the immune system is not strong enough, the mass of faulty cells will continue to grow.

Cancer, then, is not an attack from without but a breakdown
within./389-390

What I may add here to Capra's well-formulated analysis is that this 'popular
image of cancer' is not the result of folk wisdom, or folk delusion, but rather
of *folk hypnosis*. The general public knows intuitively that what the official
rhetoric says about cancer is not true, but what can they do against the
medical establishment?

Wilhelm Reich was legally murdered by his fellow colleagues, medical doctors who populate the rings of the FDA, in their effort to ruthlessly discard
out any approach and any practitioner that does not fit in their moronic,
business-driven, medical approach that serves to fill the pockets of multinationals and that has not even in a dream the interest to heal anybody. What
it does is to keep people sick because it's on the back of sick people, and
not on the back of healthy, and critical, people that it makes its return of investment. In fact, the public is brainwashed by a medical propaganda that
has no parallel in human history and that has put the image of cancer as a
murder disease in the minds of all and everybody. It's not human feeling
and natural intuition of the 'common man' that has created this standard
metaphor of the hopeless and passive patient who is 'innocently executed'
by a terminal disease. It's a myth through and through, but it can spread like
a virus because of the apathy of most consumer-citizens to see through the
veil of lies they are presented every day in the media, and it's the price they
pay for their eternal passivity to find out for themselves where the truth is.

I have an Asian friend living in the United States who is afraid to use his
Asian seasoning and incense because both is demonized in the USA as allegedly causing cancer. But that *Coca Cola* and hundreds of popular snack
food brands are highly cancerogenous, which has been proven by several
medical doctors, one of whom was attacked by Coca Cola's lawyers, but

went out of the trial unharmed, and could publish her book, you will never get to hear in any of the international media channels. It's easy to make down Asian approaches to life, and Asian products, while they are generally much more healthy than Western cuisine, when you control the media of the whole world. And it's easy to demonize them because you will ensure that these products are not going to compete in your market. I think if there was any even slight danger with using incense, major Buddhist authorities including the Dalai Lama would have been alerted. Capra continues:

Fritjof Capra

The Simontons and other researchers have developed a psycho-somatic model of cancer that shows how psychological and physical states work together in the onset of the disease. Although many details of this process still need to be clarified, it has become clear that the emotional stress has two principal effects. It suppresses the body's immune system and, at the same time, leads to hormonal imbalances that result in an increased production of abnormal cells. Thus optimal conditions for cancer growth are created. The production of malignant cells is enhanced precisely at a time when the body is least capable of destroying them. As far as the personality configuration is concerned, the individual's emotional states seem to be the crucial element in the development of cancer. The connection between cancer and emotions has been observed for hundreds of years, and today there is substantial evidence for the significance of specific emotional states. These are the result of a particular life history that seems to be characteristic of cancer patients. Psychological profiles of such patients have been established by a number of researchers, some of whom were even able to predict the incidence of cancer with remarkable accuracy on the basis of these profiles./391

When you see that emotional abuse is the most widespread and damaging abuse of all forms of abuse, which is now firmly established by the newest clinical research, then you can see why so many people get cancer in Western culture. It's because they have been emotionally distorted, if not perverted, as early as in childhood, and mainly through the rampant forced-upon co-dependence they were subjected to as children and that fixated them emotionally, sexually and physically on their parents, in a nuclear family setting, instead of experiencing natural freedom and autonomy to live their loves with peers within a protected range of extended sociability as it is, for example, still existent in Asia. It's here where the cancer etiology starts, and not later in work-stress related problems that have a much lesser impact upon our primary emotional conditioning. It's in childhood that it starts, and when the child is forbidden to be sexual, and denied to be an erotically complex human.

Published by Sirius-C Media Galaxy LLC, 2011

Fritjof Capra

The Web of Life

A New Scientific Understanding of Living Systems
New York: Anchor Books, 1997

The Web of Life is perhaps Capra's best and most important book. Why, one may want to ask? Because it was in this book that Capra really defined his approach to ecology, thereby making ecology, or deep ecology, a concept that is part of a new science paradigm, powerfully introduced and promoted by one of the most important science theorists of our times. What is *deep ecology?* Capra writes:

> **Fritjof Capra**
>
> Whereas the old paradigm is based on anthropocentric (human-centered) values, deep ecology is grounded in ecocentric (earth-centered) values. It is a worldview that acknowledges the inherent value of nonhuman life./11

And why do we need it?

> **Fritjof Capra**
>
> Such a deep ecological ethics is urgently needed today, and especially in science, since most of what scientists do is not life-furthering and life-preserving but life-destroying. With physicists designing weapon systems that threaten to wipe out life on the planet, with chemists contaminating the global environment, with biologists releasing new and unknown types of microorganisms without knowing the consequences, with psychologists and other scientists torturing animals in the name of scientific progress – with all these activities going on, it seems most urgent to introduce 'ecoethical' standards into science./11

This book's quest is enormous, in that it requires modern science to fundamentally shift its regard upon nature, and upon living. Our regard upon nature has been conditioned by patriarchy since about five thousand years, and it's a rather defensive, distorted, if not completely schizophrenic regard, so much the more as both our mainstream religious paradigm and the *Cartesian* shift of science in the 17th and 18th centuries have contributed heavily to this reductionist view of nature. Capra looked back in history and found amazing early intuitions and truths propagated by our great thinkers, poets and philosophers, such as for example *Immanuel Kant, Johann Wolfgang von Goethe* or *William Blake*. He writes:

Fritjof Capra

The understanding of organic form also played an important role in the philosophy of Immanuel Kant, who is often considered the greatest of the modern philosophers. An idealist, Kant separated the phenomenal world from a world of 'things-in-themselves'. He believed that science could offer only mechanical explanations, but he affirmed that in areas where such explanations were inadequate, scientific knowledge needed to be supplemented by considering nature as being purposeful./21

On the same line of thinking, Capra investigated what the earth, the globe, the planet means for us today, and why our science and technologies are so deeply hostile to it and so little caring for its preservation? He found conclusive answers in ancient traditions that fostered what we call today a *Gaia* worldview, a respectful attitude toward the earth, the mother, the yin energy and generally female values:

Fritjof Capra

The view of the Earth as being alive, of course, has a long tradition. Mythical images of the Earth Mother are among the oldest in human religious history. Gaia, the Earth Goddess, was revered as the supreme deity in early, pre-Hellenic Greece. Earlier still,

Published by Sirius-C Media Galaxy LLC, 2011

from the Neolithic through the Bronze Ages, the societies of 'Old Europe' worshiped numerous female deities as incarnations of Mother Earth./22

This is how Capra, always grounded in common sense and meaningful retrospection smoothly introduces the novice reader to the concept of *systems research* or the *systems worldview*. Historically we can observe a certain evolution in post-matriarchal thought, which was naturally systemic, from the 'Atomistic Worldview' *(Democritus)*, over the 'Cartesian Worldview' *(Newton, La Mettrie, René Descartes)* and the 'Relativistic Worldview' *(Einstein, Planck, Heisenberg)*, to the 'Systemic Worldview' *(Bohm, Bateson, Grof, Capra, Laszlo, etc.)* and the 'Holistic Worldview' *(Talbot, Goswami, McTaggart, etc.)*. In all systems, we have to deal with different levels of complexity that are woven in each other, thus rendering it almost impossible to dissect parts of the system for closer research without distorting the whole of our research results.

This means that, contrary to earlier vivisectionist science, we have to leave the system intact and focus our research onto the whole of it - which makes our research very complex by definition. Hence, we had to develop a new mathematics, which today is called the mathematics of complexity, in order to deal with the high complexity levels in living systems. This also means that our usual way of analysis as a scientific method did not work any more for our inquiry in the functionality of living systems. Capra explains:

Fritjof Capra

According to the systems view, the essential properties of an organism, or living system, are properties of the whole, which none of the parts have. They arise from the interactions and relationships among the parts. These properties are destroyed when the system is dissected, either physically or theoretically, into isolated elements. Although we can discern individual parts in any system, these parts are not isolated, and the nature of the whole is always different from the mere sum of its parts. (...) The great shock of twentieth-century science has been that systems cannot be understood by analysis. The properties of the parts are not intrinsic properties but can be understood only within the context of the larger whole. Thus the relationship between the parts and the whole has been reversed./29

Capra defines the *Web of Life* as networks within networks.

Fritjof Capra

At each scale, under closer scrutiny, the nodes of the network reveal themselves as smaller networks. We tend to arrange these systems, all nesting within larger systems, in a hierarchical scheme by placing the larger systems above the smaller ones in pyramid fashion. But this is a human projection. In nature there is no 'above' or 'below', and there are no hierarchies. There are only networks nesting within other networks./35

In fact, living systems are not, as most of our governmental and societal organization, hierarchical, but network-based, and thus expanding not up-to-down but horizontally by 'neuronally' linking segments to larger molecular structures that distribute information with the speed of the light over the whole of the network. You can also say that a living network is a system of *total information sharing* where there is not one single molecule that is uninformed at any point in time and space. The fact that horizontal networks are nested within other horizontal networks, while the different networks all possess a different level of complexity, makes research so intricate. This is why high-performance computers have greatly aided in developing systems theory.

Fritjof Capra

In the new systems thinking, the metaphor of knowledge as a building is being replaced by that of the network. As we perceive reality as a network of relationships, our descriptions, too, form an interconnected network of concepts and models in which

Published by Sirius-C Media Galaxy LLC, 2011

there are no foundations. (…) When this approach is applied to science as a whole, it implies that physics can no longer be seen as the most fundamental level of science. Since there are no foundations in the network, the phenomena described by physics are not any more fundamental than those described by, say, biology or psychology. They belong to different systems levels, but none of those levels is any more fundamental than the others./39

But the most revolutionary insight here is that our usual habit of dissecting parts of a whole for further scrutiny and scientific investigation does not work any more with living systems. Why is this so?

Fritjof Capra

Ultimately – as quantum physics showed so dramatically – there are no parts at all. What we call a part if merely a pattern in an inseparable web of relationships. Therefore the shift from the parts to the whole can also be seen as a shift from objects to relationships./37

Hence, the whole of our approach to scientific investigation has to shift from an object-based to a *relationship-based research approach* when we deal with living systems. This requires the researcher to change his inner setup; this is what quantum physics revealed to us: the observer's belief system will be reflected in the outcome of the research, as it is part of reality, and not to be dissected from it.

Fritjof Capra

When we draw a picture of a tree, most of us will not draw the roots. Yet the roots of a tree are often as expansive as the parts we see. In a forest, moreover, the roots of all trees are interconnected and form a dense underground network in which there are no precise boundaries between individual trees. In short, what we call a tree depends on our perceptions. It depends, as we say in science, on our methods of observation and measurement. In the words of Heisenberg: 'What we observe is not nature itself, but nature exposed to our method of questioning.' Thus systems thinking involves a shift from objective to 'epistemic' science, to a framework in which epistemology – 'the method of questioning' – becomes an integral part of scientific theories./40

And there is one more crucial element in systems research that Capra explains and elucidates. It is what we already learnt within the revolutionary reframing of Western science by quantum physics, the fact namely that in approaching quantum reality, and organic behavior, we have to learn the mathematics of probability.

Fritjof Capra

The new mathematics … is one of relationships and patterns. It is qualitative rather than quantitative and thus embodies the shift of emphasis that is characteristic of systems thinking – from objects to relationships, from quantity to quality, from substance to pattern./113

What is probability? It is the approximation of behavior. Dealing with approximations means that we leave the certainty principle and venture into what *Heisenberg* called the *uncertainty principle*. Giving up certainty triggers fear. And this fear was very vividly described by *Max Planck* and *Heisenberg* when the paradigm began to shift and quantum physics slowly but definitely began to undermine Euclidian geometry and Newtonian assuredness. Why has our certainty about the universe been undermined? Well, when we look at Hindu philosophy and ancient Chinese science, certainty was actually never an element of holistic science, but only a part of fragmented science. When we abandon certainty, we begin to grasp the notions of *approximation*, and of *probability*, and accordingly we will shift our mathematical constructs:

> **Fritjof Capra**
>
> What makes it possible to turn the systems approach into a science is the discovery that there is approximate knowledge. This insight is crucial to all of modern science. The old paradigm is based on the Cartesian belief in the certainty of scientific knowledge. In the new paradigm it is recognized that all scientific concepts and theories are limited and approximate. Science can never provide any complete and definite understanding./41

The next important centerpoint in the *Web of Life* is the introduction of the notion of open systems. What is an open system? Capra explains:

> **Fritjof Capra**
>
> Unlike closed systems, which settle into a state of thermal equilibrium, open systems maintain themselves far from equilibrium in this 'steady state' characterized by continual flow and change./48

Living systems are open systems, and not closed systems which means that their main characteristic is *change and flow*, and not continuity and static behavior. And they are far from equilibrium, which is the single most revolutionary discovery of systems research. Which means living systems are constantly struggling against decay. And decay here means equilibrium. This is a very important insight as when we extrapolate this insight from organic systems into our metaphysical reality, we see that it applies also to human beings, and even to religions. When we are settled, we are dead. This is what it all boils down to. And this insight from systems research may help us to survive in a state far from equilibrium, putting our assuredness or fake assuredness away, to stay with probability, the eternal beginner's mind, as it is so wistfully expressed in *Zen*.

I have stressed in all my publications the importance of understanding the nature of our universe as a basically *patterned universe*, stressing the importance in nature of patterned intelligence, or patterned organization. What are patterns? Capra explains the importance of pattern when he explores the meaning of *self-organization*, which is one major characteristic of living systems:

Fritjof Capra

To understand the phenomenon of self-organization, we first need to understand the importance of pattern. The idea of a pattern of organization – a configuration of relationships characteristic of a particular system – became the explicit focus of systems thinking in cybernetics and has been a crucial concept ever since. From the systems point of view, the understanding of life begins with the understanding of pattern./80

In order to scientifically explain patterns we need to change or for the least upgrade our basic toolset of scientific investigation. Capra explains:

Fritjof Capra

In the study of structure we measure and weigh things. Patterns, however, cannot be measured or weighed; they must be mapped. To understand a pattern we must map a configuration of relationships. In other words, structure involves quantities, while pattern involves qualities./81

This really involves a radical change in our scientific thinking because traditionally Cartesian science was quantity-based and measure-oriented, while systemic science is quality-based and relationship-oriented, a truth that Capra exemplifies when looking at the properties involved in the scientific focus of both static and systemic science theory:

Fritjof Capra

Systemic properties are properties of pattern. What is destroyed

Published by Sirius-C Media Galaxy LLC, 2011

when a living organism is dissected is its pattern. The components are still there, but the configuration of relationships among them – the pattern – is destroyed, and thus the organism dies./81

An important self-regulatory function in living systems are *feedback loops*. Without feedback loops, living systems could not be self-organizing. Capra explains:

Fritjof Capra

Because networks of communication may generate feedback loops, they may acquire the ability to regulate themselves. For example, a community that maintains an active network of communication will learn from its mistakes, because the consequences of a mistake will spread through the network and return to the / source along feedback loops. Thus the community can correct its mistakes, regulate itself, and organize itself. Indeed, self-organization has emerged as perhaps the central concept in the systems view of life, and like the concepts of feedback and self-regulation, it is linked closely to networks. The pattern of life, we might say, is a network pattern capable of self-organization. This is a simple definition, yet it is based on recent discoveries at the very forefront of science./82-83

Another centerpoint in this book is Capra's focus upon the intrinsic quality of living systems as *nonlinear systems* that require, to be understood, an equally *nonlinear* mathematical approach:

Fritjof Capra

In nonlinear systems … small changes may have dramatic effects because they may be amplified repeatedly by self-reinforcing feedback. Such nonlinear feedback processes are the basis of the instabilities and the sudden emergence of new forms of order that are so characteristic of self-organization./124

One early realization of mathematical nonlinearity was the introduction of the fractal in mathematics. In fact, in my exchanges with the Swiss mathematician *Peter Meyer* who was the collaborator of *Terence McKenna* for the realization of the *Timewave Zero* calculus as a part of *Novelty Theory*, I learnt that time is a fractal. Capra explains:

Fritjof Capra

The great fascination exerted by chaos theory and fractal geometry on people in all disciplines – from scientists to managers to artists – may indeed be a hopeful sign that the isolation of mathematics is ending. Today the new mathematics of complexity is / making more and more people realize that mathematics is much more than dry formulas; that the understanding of pattern is crucial to understand the living world around us; and that all questions of pattern, order, and complexity are essentially mathematical./152-153

After having elucidated that systems research involves a process-based scientific approach rather than an object-based one, Capra presents the perhaps most important research topic in this book: the reinvestigation of *cognition* based on the insights from systems research. Capra pursues:

Fritjof Capra

The identification of mind, or cognition, with the process of life is a radically new idea in science, but it is also one of the deepest and most archaic intuitions of humanity. In ancient times the rational human mind was seen as merely one aspect of the immaterial soul, or spirit./264

In fact, the whole debate about information processing, vividly criticized in the early writings of think tank *Edward de Bono,* and the even larger debate about cybernetics make it all clear that cognition is currently in a process of profound reevaluation:

Published by Sirius-C Media Galaxy LLC, 2011

Fritjof Capra

The computer model of cognition was finally subjected to serious questioning in the 1970's when the concept of self-organization emerged. (…) These observations suggested a shift of focus – from symbols to connectivity, from local rules to global coherence, from information processing to the emergent properties of neural networks./266

In my scientific exploration of emotions, and ▸pedoemotions, I have revisited our scientific grasp of emotions, as it was cognized within a fragmented and reductionist manner under the *Cartesian* science paradigm. Capra comprehensively explains that emotions are not singular elements but coherently organized within a *patterned system* in which cognition and response are intertwined in a self-regulatory and organic whole:

Fritjof Capra

The range of interactions a living system can have with its environment defines its 'cognitive domain'. Emotions are an integral part of this domain. For example, when we respond to an insult by getting angry, that entire pattern of physiological processes – a red face, faster breathing, trembling, and so on – is part of cognition. In fact, recent research strongly indicates that there is an emotional coloring to every cognitive act./269

The most important fact that systems theory teaches us about cognition is that it does not at all work like a computer processes information. Information processing, already years ago in the words of *Edward de Bono* has been called a preoccupation of Western scientists, and this obsession was not justified because our brain *does not process* information as a computer does. Capra explains why:

Fritjof Capra

A computer processes information, which means that it manipulates symbols based on certain rules. The symbols are distinct elements fed into the computer from outside, and during the information processing there is no change in the structure of the machine. The physical structure of the computer is fixed, determined by its design and construction. The nervous system of a living organism … interacts with its environment by continually modulating its structure, so that at any moment its physical / structure is a record of previous structural changes. The nervous system does not process information from the outside world but, on the contrary, brings forth a world in the process of cognition./274-275

Capra then answers to the debate about *artificial intelligence* and the myths it creates in the minds of masses of people:

Fritjof Capra

A lot of confusion is caused by the fact that computer scientists use words such as intelligence, memory, and language to describe computers, thus implying that these expressions refer to the human phenomena we know well from experience. This is a serious misunderstanding. For example, the very essence of intelligence is to act appropriately when a problem is not clearly defined and solutions are not evident. Intelligent human behavior in such situations is based on common sense, accumulated from lived experience. Common sense, however, is not available to computers because of their blindness of abstraction and the intrinsic limitations / of formal operations, and therefore it is impossible to program computers to be intelligent./275-276

In my exchanges with young men, most of whom are computer programmers and IT technicians, I have been stunned over years about their lack of true intelligence, their robotic, lifeless mechanical thinking and their almost total lack of empathy. These men repeated formulas in their minds when confronted with any given problem, and they lacked the most basic creativity to find solutions for problems that are novel.

One of these men wrote me his IQ had once been tested as 181, and he indeed performed well as a computer programmer. But when it came to solve the simple problems of his love life, he was completely stuck. His intelligence knew to handle computers, but it was at pains to understand the most basic forms of human behavior, and especially, emotional reactions. Emotions were for him a matter of anxiety and confusion. Not surprisingly so, he and most of his friends fostered the view that humans should always be *cool and rational* and that irrationality should be *rooted out*. And they were not aware how irrational they were in their obsession to be rational! And their main occupation, you guess it, was studying artificial intelligence. Real intelligence is human, and original, not mechanical, and artificial! Artificial intelligence is artificial stupidity!

True intelligence is contextual, as language is. No computer can understand meaning. A rat's intelligence is a million times closer to that of man than that of the most powerful and sophisticated computer. Capra notes:

> **Fritjof Capra**
>
> The reason is that language is embedded in a web of social and cultural conventions that provides an unspoken context of meaning. We understand this context because it is common sense to us, but a computer cannot be programmed with common sense and therefore does not understand language./276

The results we see every day on the Internet. I once was in exchange with one of those who were working on the Systran translation system. He was sitting every day ten hours in front of nine computers and firmly believed that in about ten years he would have bred the perfect computer translator! He had not the faintest idea that computers are basically stupid and that all translations are approximations that require contextual intelligence - which a computer does not have, and will never have, do what you will. Capra pursues:

Fritjof Capra

Mind is not a thing but a process – the process of cognition, which is identified with the process of life. The brain is a specific structure through which this process operates. Thus the relationship between mind and brain is one between process and structure./278

Truly intelligent people are not interested in computers, or not more than people are interested in their cooking pan in the kitchen. The cooking pan as the computer fulfill needs; they are not companions! Intelligent people are interested in *ecology*. It's ecology that is needed in our thinking and in our world because it has been shunned throughout patriarchy, and if we are not going to implement a real *science of ecology*, we will be wiped out as a human race at some point in the future.

Capra is known to be one of the finest ecologists around, and he is very often in Germany for giving ecological advice to the German government and non-governmental agencies. He has put a stress on *sustainability*, a term that was introduced in the early 1980s by Lester Brown, founder of the *Worldwatch Institute*. He defined a sustainable society as one that is able to satisfy its needs without diminishing the chances of future generations. Thus, a system is sustainable when it's not only functional but also well integrated in a greater continuum so that it has a good prognosis for survival, for continuity. Capra writes:

Fritjof Capra

Partnership is an essential characteristic of sustainable communities. The cyclical exchanges of energy and resources in an ecosystem are sustained by pervasive cooperation. Indeed, we have seen that since the creation of the first nucleated cells over two billion years ago, life on Earth has proceeded through ever more intricate arrangements of cooperation and coevolution. Partnership – the tendency to associate, establish links, live inside one another, and cooperate – is one of the hallmarks of life./301

Partnership and cooperation were indeed alien words under patriarchy but they were imbedded in the pre-patriarchal civilizations, such as the Minoan Civilization, and thus what we get today is a return to the sources. Unfortunately most of our governments today have a rather cynical attitude when it goes to recognize the need to protect our earth from being destroyed by our ruthless and non-ecological technologies. Capra is quite outspoken here:

Fritjof Capra

The 1991 war in the Persian Gulf, for / example, which killed hundreds of thousands, impoverished millions, and caused unprecedented environmental disasters, had its roots to a large extent in the misguided energy policies of the Reagan and Bush administrations./299-300

And hopefully so, the fact that dedicated ecologists such as Fritjof Capra are today traveling the world to consult governments in drafting more ecologically friendly policies will contribute to change both our administrative and business thinking and put it on the brave new ground of sound ecology, partnership, cooperation and respectful communication that crosses national and cultural borders. And this is what Capra says about this subject, in his book *The Hidden Connections (2002),* p. 99:

Fritjof Capra

Organizations need to undergo fundamental changes, both in order to adapt to the new business environment and to become ecologically sustainable. This double challenge is urgent and real, and the recent extensive discussions of organizational change are fully justified. However, despite these discussions and some anecdotal evidence of successful attempts to transform organizations, the overall track record is very poor.

Published by Sirius-C Media Galaxy LLC, 2011

Fritjof Capra

The Hidden Connections

Integrating the Biological, Cognitive, and Social Dimensions of Life into a Science of Sustainability, New York: Anchor Books, 2002

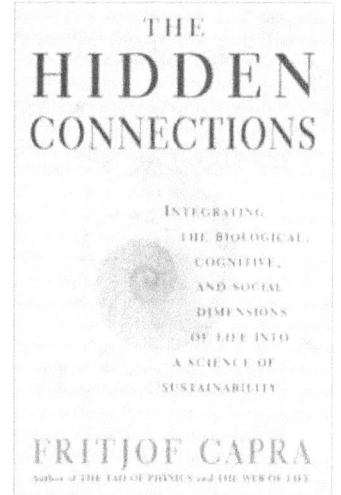

In my personal opinion, *The Hidden Connections* is the most lucid of Capra's five books reviewed in this book. My opinion is based on a sequential reading of all of Capra's books in that I started with the earliest and finished with the last. I read the first two, the *Tao of Physics* and *The Turning Point* around twenty years ago, and the last three only two years ago! So I do not guarantee that you will have the same impression if you don't follow the sequential order, and give yourself time to ponder deeply about the content. You may find I exaggerate, but truly, I needed twenty years to really digest the content Capra presented, in all detail, in *The Turning Point*. I have read the book three times, and every time I got to see new details. For in a way, Capra's books are *educational*, as it were, because they teach us many things we need to learn for our transiting into a more ecologically wise era; and I do not know any other author who got so much of this special talent to explain complex things in an easy and comprehensive manner.

This being said, I could well imagine that if you begin reading Capra with the present book, you might get stuck somewhere somehow in the midst of it - simply because you lack out on essential information that is contained in Capra's earlier books. That's how I see that. So if you want to make a rushy bedtime lecture from this book, I rather disadvise you this kind of diet. Better you begin with the beginning. Read the *Tao of Physics* first, then *The*

Turning Point, then read *The Web of Life*, and then the present book. Then you will be guided!

At the very start of *The Hidden Connections*, and yes, many hidden things will indeed be revealed in this book. Capra communicates an important detail about himself and his unusual development as a scientist:

Fritjof Capra

My extension of the systems approach to the social domain explicitly includes the material world. This is unusual, because traditionally social scientists have not been very interested in the world of matter. Our academic disciplines have been organized in such a way that the natural sciences deal with material structures while the social sciences deal with social structures, which are understood to be, essentially, rules of behavior. In the future, this strict division will no longer be possible, because the key challenge of this new century – for social scientists, natural scientists and everybody else – will be to build ecologically sustainable communities, designed in such a way that their technologies and social institutions – their material and social structures – do not interfere with nature's inherent ability to sustain life./xix

Capra starts, systemically sound, with the cell, noting that the simplest living system is the cell, and especially, the bacterial cell. Then Capra looks at what *membranes* are, and what they do, and this teaches us an important lesson about relationships. I haven't found this insightful metaphor anywhere else, and it showed me right at the start of this book that it's going to be substantial reading:

Fritjof Capra

A membrane is very different from a cell wall. Whereas cell walls are rigid structures, membranes are always active, opening and closing continually, keeping certain substances out and letting others in. The cell's metabolic reactions involve a variety of ions, and the membrane, by being semipermeable, controls their proportions and keeps them in balance. Another critical activity of the membrane is to continually pump out excessive calcium waste, so that the calcium remaining within the cell is kept at the precise, very low level required for its metabolic functions. All these activities help to maintain the cell as a distinct entity and protect it from harmful environmental influences. Indeed, the first thing a bacterium does when it is attacked by another organism is to make membranes./8

The next important point to understand how nature 'thinks' is the cell's metabolism, the network that serves recycling. Capra succinctly elaborates:

Fritjof Capra

When we take a closer look at the processes of metabolism, we notice that they form a chemical network. This is another fundamental feature of life. As ecosystems are understood in terms of food webs (networks of organisms), so organisms are viewed as networks of cells, organs and organ systems, and cells as networks of molecules. One of the key insights of the systems approach has been the realization that the network is a pattern that is common to all life. Wherever we see life, we see networks. (...) The metabolic network of a cell involves very special dynamics that differ strikingly from the cell's nonliving environment. Taking in nutrients from the outside world, the cell sustains itself by means of a network of chemical reactions that take place inside the boundary and produce all of the cell's components, including those of the boundary itself./9

I shall leave out in this review the long passages in which Capra explains the essential contributions of systems researchers such as Varela, Maturana or Prigogine, as this would render this review definitely too extensive, and restrict myself on a few remarks to describe the core of systems research that Capra unfolds in this book:

Fritjof Capra

The starting point for this is the observation that all cellular structures exist far from equilibrium state – in other words, the cell would die – if the cellular metabolism did not use a continual flow of energy to restore structures as fast as they are decaying. This means that we need to describe the cell as an open system. Living systems are organizationally closed – they are autopoietic networks – but materially and energetically open./13

One of the most important insights we gain from systems theory and the close observation of natural processes are focusing on the topics of chaos and order. What is chaos? What is order? We all have some preconceptions here. Sure, but I promise you that when you read this book, you will let them all go, simply because they are wrong! Chaos is not chaos, but ordered chaos, and thus not just random. Besides, order, real order, is not a natural condition! Yes, order is not good. I mean it's not a stable condition. And if it is, you are dead! You see that? Well, you may remember that we have briefly discussed already in the review of *The Web of Life* what self-organization means relating to systems. Here, Capra explains in more detail what self-organization actually does:

Fritjof Capra

Th[e] spontaneous emergence of order at critical points of instability is one of the most important concepts of the new understanding of life. It is technically known as self-organization and is often referred to simply as emergence. It has been recognized as the dynamic origin of development, learning and evolution. In other words, creativity – the generation of new forms – is a key property of all living systems. And since emergence is an integral part of the dynamics of open systems, we reach the important conclusion that open systems develop and evolve. Life constantly reaches out into novelty./14

Published by Sirius-C Media Galaxy LLC, 2011

The next great error most of us were caught in and that was the result of the left-brain hypertrophy was the distinction we have taken between humans and animals when it was about cognition. Not so. We are not much more intelligent than Gorillas, only a little more, to be precise only a factor of 1.6 more. And besides that, it was believed that in animals cognition was working in basically different ways than in humans. This seems to be an error, as researchers found you can talk with chimpanzees if you learn their language, and they can learn yours. Capra summarizes this research shortly:

Fritjof Capra

The unified, post-Cartesian view of mind, matter, and life also implies a radical reassessment of the relationships between humans and animals. Throughout most of Western philosophy, the capacity to reason was seen as a uniquely human characteristic, distinguishing us from all other animals. The communication studies with chimpanzees / have exposed the fallacy of this belief in the most dramatic of ways. They make it clear that the cognitive and emotional lives of animals and humans differ only by degree; that life is a great continuum in which differences between species are gradual and evolutionary./65-66

I shall finalize this book review with some very interesting political and social hidden connections that Capra unveils in his book. There are probably still people around who are fond of biotechnology, but I guess they just ignore the facts, and their knowledge is for the most part taken from the huge amount of propaganda material around in popular science magazines and web sites. And was it only for this enlightening information, this book is worth its price because it daringly unveils the hidden facts and tells the truth!

Fritjof Capra

The most widespread use of plant biotechnology has been to develop herbicide-tolerant crops in order to boast the sales of particular herbicides. There is a strong likelihood that the transgenic plants will cross-pollinate with wild relatives in their surroundings, thus creating herbicide-resistant superweeds. Evidence indicates that such gene flows between transgenic crops and wild relatives are already occurring./193

Why do we need biotechnology, if I may ask? I guess we don't need it, but certain people and their consorts need it for making huge amounts of money. So is that democracy? Is it typical for a democracy that all suffer from the side effects of technologies that enrich a few? Is it democracy that favors paradigms and social policies that bring damage to our planet? I learnt at law school that that kind of system is called an *oligarchy*, the reign of a few. So how did we ever come to say that we are living in a democracy?

Fritjof Capra

In the animal kingdom, where cellular complexity is much higher, the side effects in genetically modified species are much worse. 'Super-salmon' which were engineered to grow as fast as possible, ended up with monstrous heads and died from not being able to breathe or feed properly. Similarly, a superpig with a human gene for a growth hormone turned out ulcerous, blind, and impotent. (…) The most horrifying and by now best-known story is probably that of the genetically altered hormone called recombinant bovine growth hormone, which has been used to stimulate milk production in cows despite the fact that American dairy farmers have produced vastly more milk than people can consume for the past fifty years. The effects of this genetic engineering folly on the cow's health are serious. They include bloat, diarrhea, diseases of the knees and feet, cystic ovaries, and many more. Besides, their milk may contain a substance that has been implicated in human breast and stomach cancers./198

Why do we need superpigs? It seems to me that they are the result of *quantitative* thinking, of a primacy of quantity over quality, and this for the obvi-

Published by Sirius-C Media Galaxy LLC, 2011

ous reason of maximizing profits. This is a good example for the fact that we live in what has been called *The Corporate Society* or *Corporate America*, as the prototype of a society in which major corporations dictate the standards the government is going to follow and to enact as laws. Capra notes the details:

Fritjof Capra

In the United States, the biotech industry has persuaded the Food and Drug Administration (FDA) to treat GM food as substantially equivalent to traditional food, which allows food producers to evade normal testing by the FDA and the Environmental Protection Agency (EPA), and also leaves it to the companies' own discretion as to whether to label their products as genetically modified. Thus, the public is kept unaware of the rapid spread of transgenic foods and scientists will find it much harder to trace harmful effects. Indeed, buying organic is now the only way to avoid GM foods./199

In Germany and France, the laws are different regarding genetically modified food and the European Community will probably ban all products that are to be subsumed under this term, because this is already the state of the law in Germany and France, for good reasons. Capra informs:

Fritjof Capra

The governments of France, Italy, Greece, and Denmark announced that they would block the approval of new GM crops in the European Union. The European Commission made the labeling of GM foods mandatory, as did the governments of Japan, South Korea, Australia, and Mexico. In January 2000, 130 nations signed the groundbreaking Cartagena Protocol on Biosafety in Montreal, which gives nations the right to refuse entry to any genetically modified forms of life, despite vehement opposition from the United States./228

As a lawyer, I can clearly see that we are facing currently a challenge to legally codify these new technologies - or, as it were, they are going to codify us, entraining us in a turbulence of faits établis, and then the law will leap behind the actual developments. But the law should better accompany the research step by step so as to be updated with the explosive growth of these very heavily funded research disciplines. Capra writes:

Fritjof Capra

The development of such new biotechnologies will be a tremendous intellectual challenge, because we still do not understand how nature developed technologies during billions of years of evolution that are far superior to our human designs. How do mussels produce glue that sticks to anything in water? How do spiders spin a silk thread that, ounce for ounce, is five times stronger than steel? How do abalone grow a shell that is twice as tough as our high-tech ceramics? How do these creatures manufacture their miracle materials in water, at room temperature, silently, and without any toxic byproducts?/204

I suggest that the intellectual and political challenge will perhaps not be the development of such new biotechnologies, but rather the stop of their development. When I say this I mean we should carefully restrict their use by law. We should eventually begin to understand that our intelligence simply is not on the same level as that of Mother Nature, and playing around with that kind of stuff truly is not child play.

Published by Sirius-C Media Galaxy LLC, 2011

Fritjof Capra, Gunter Pauli, Eds.

Steering Business Toward Sustainability

New York: United Nations University Press, 1995

Steering Business Toward Sustainability is a book of high practical value for leaders and organizations who are conscious of the need for deep ecology and the challenge we presently face to update most of our basic business and investment routines and procedures in order to build not destructive and depletive, but sustainable organizations. Capra is straightforward:

> **Fritjof Capra**
>
> Quite simply, our business practices are destroying life on earth. Given current corporate practices, not one wildlife reserve, wilderness, or indigenous culture will survive the global market economy./1

Capra's idea of deep ecology has grown over many years. It is rooted in the insights he exposed in his previous four books, and thus we can say this is solidly grounded. In addition, Capra leaves no doubt that it's not just a technocratic idea, but an intrinsically spiritual concept. He also credits those, religions and peoples, who have practiced ecological thinking long before the birth of the United States of America:

> **Fritjof Capra**
>
> When the concept of the human spirit is understood as the mode of consciousness in which the individual feels connected to the cosmos as a whole, it becomes clear that ecological awareness is spiritual in its deepest essence. It is therefore not surprising that the emerging new vision of reality, based on deep ecological awareness, is consistent with the so-called perennial philosophy of spiritual traditions, whether we talk about the

spirituality of Christian mystics, that of Buddhists, or the philosophy and cosmology underlying the American Indian traditions./3

Capra reminds us of the fact that when restructuring our economies, we should learn from nature, instead of feeling superior over nature. *Ecological literacy* is one of the notions Capra is currently lecturing about in Germany, and many other countries, and Gunter Pauli, the co-editor of this reader is one of Capra's truest collaborators, and himself an authority on ecology in Germany. Within the concept of ecological literacy, Capra seems to give the highest importance to the term *sustainability*, and comprehensively explains what this term means:

Fritjof Capra

In our attempts to build and nurture sustainable communities we can learn valuable lessons from ecosystems, because ecosystems are sustainable communities of plants, animals, and microorganisms. To understand these lessons, we need to learn nature's language. We need to become ecologically literate. (...) Being ecologically literate means understanding how ecosystems organize themselves so as to maximize sustainability./4

Many of us have not yet understood why our modern technologies are so much in conflict with nature's setup, and this is a fact that is hardly ever elucidated in the mass media. Non-educated people, and even entrepreneurs who have not been exposed to academic study are usually at pains with understanding the deeper reasons of this conflict. Capra, referencing Paul Hawken, *The Ecology of Commerce*, New York: Harper & Row, 1993, elucidates it:

Fritjof Capra

The present clash between business and nature, between economics and ecology, is mainly due to the fact that nature is cyclical, whereas our industrial systems are linear, taking up energy and resources from the earth, transforming them into products plus waste, discarding the waste, and finally throwing away also after they have been used. Sustainable patterns of production and consumption need to be cyclical, imitating the processes in ecosystems./5

Back in Antiquity, there was hardly a need for people to learn systems thinking because they were naturally aligned with the logic of nature, because they were living with nature, and not on top of nature, as we do today. We can also say that we as modern city dwellers have lost our continuum, as it was expressed with much emphasis by Jean Liedloff in *The Continuum Concept (1977/1986)*.

Besides, Capra informs us about how we should apply ecology in our daily lives, and what it teaches us. There are seven principles to learn that Capra calls *Principles of Ecology* and that he explains one by one:

Fritjof Capra

Interdependence
All members of an ecosystem are interconnected in a web of relationships, in which all life processes depend on one another.

Ecological Cycles
The interdependencies among the members of an ecosystem involve the exchange of energy and resources in continual cycles.

Energy Flow
Solar Energy, transformed into chemical energy by the photosynthesis of green plants, drives all ecological cycles.

Partnership
All living members of an ecosystem are engaged in a subtle in-

terplay of competition and cooperation, involving countless forms of partnership.

Flexibility
Ecological cycles have the tendency to maintain themselves in a flexible state, characterized by interdependent fluctuations of their variables.

Diversity
The stability of an ecosystem depends on the degree of complexity of its network of relationships; in other words, on the diversity of the ecosystem.

Coevolution
Most species in an ecosystem coevolve through an interplay of creation and mutual adaptation.

Sustainability
The long-term survival of each species in an ecosystem depends on a limited resource base. Ecosystems organize themselves according to the principles summarized above so as to maintain sustainability./6

Capra also explains very well the feedback-looping that we find is a typical feature of living systems. The understanding of feedbacking as constant parameter change as a response to a given stimulus is crucial for the understanding of the cyclic nature of all life. This is one of the points modern scientists are really at pains with because their thought structure simply is too linear. Capra explains:

Fritjof Capra
When changing environmental conditions disturb one link in an ecological cycle, the entire cycle acts as a self-regulating feedback loop and soon brings the situation back into balance. And

Published by Sirius-C Media Galaxy LLC, 2011

since these disturbances happen all the time, the variables in an ecological cycle fluctuate continually. These fluctuations represent the ecosystem's flexibility. Lack of flexibility manifests itself as stress. In particular, stress will occur when one or more variables of the system are pushed to their extreme values, which induces increased rigidity throughout the system. Temporary stress is an essential aspect of life, but prolonged stress is harmful and destructive to the system./7

On the other hand, it's exactly this widely unpredictable feedback-looping that is inherent in the current paradigm of ecological destruction. This dangerous situation is worsened by the general lack of ecological literacy about the possible effects of large disturbances, such as ozone hole, deforestation, global warming and desertification. Our knowledge also is insufficient to make ecological solutions work *effectively* even once ecology-friendly policies are implemented by governments and companies. It's not enough to see the dangers and implement good new laws for protecting our nature, we also need to understand how the damages that are already done will interact with our new policies; this is so because it's not taken for granted that our best-intended tactics of healing nature are really healing nature. For insuring this, we have to learn much more about feedback-looping in natural systems.

We have to understand that nature heals herself and that we only need to remove the factors that cause the damage. For example, it has been shown that the planting of new trees does *not* per see heal the damage that deforestation has done to ecological saneness of our planet. It's all in the why and how of planting trees, where, how many, and in what mixture of species that the wisdom lies. On the other hand, it has been seen in Indonesia, which is one of the worst hit countries by deforestation, that huge areas that were deforested began to grow trees without anybody doing anything about it! Later research showed that the conditions had been ideal for trees to grow again, but nobody really knew why at other places, where at first sight conditions were very similar, this was not the case.

We definitely have to develop humility in the face of the dreadful ignorance we have about the complexity level of nature, at all levels of evolution. We are simply not trained in complexity thinking, and our schools and universities destroy the little of complexity we have developed naturally as children as a result of free play. It is *freedom* that is at the basis of building complexity, not discipline, it is permissiveness, not repression. Here is where our morality clearly stares grimly in nature's face because nature is immoral. So we should do away with our projections upon nature and at the same time get all our senses and our emotional intelligence ready for receiving the messages of nature. Nature communicates when we are ready to listen, and it will tell us how we can help healing the damage we have done over five thousand years of patriarchal ignorance.

This book together with *Hidden Connections* and *The Web of Life* teaches the basics of understanding nature's complexity. It also teaches us the importance of *diversity*, a concept that at present is rather shunned by mainstream politics, while liberal phases, as it was the case through the 1970s, foster higher levels of cultural diversity. Nature shows us that this is not just a random development but that it's diversity on which side is intelligent and life-fostering behavior, and not uniformity. This is so, inter alia, because diversity fosters flexibility, and vice versa, while uniformity entails rigidity. Capra elucidates:

Fritjof Capra

In ecosystems, flexibility through fluctuations does not always work, because there can be very severe disturbances that actually wipe out an entire species. In other words, one of the links in the ecosystem's network is destroyed. An ecological community will be resilient when this link is not the only one of its kind;

Published by Sirius-C Media Galaxy LLC, 2011

when there are other connections that can at least partially fulfill its functions. In other words, the more complex the network, the greater the diversity of its interconnections, the more resilient it will be. The same is true in human communities. Diversity means many different relationships, many different approaches to the same problem. A diverse community is a resilient community, capable of adapting easily to changing situations./8

What does loss of diversity on the planet, in all layers of living systems, mean for our future? The regard here is rather dim, and Capra leaves no doubt about this:

Fritjof Capra

The loss of biodiversity, i.e. the daily loss of species, is in the long run one of our most several global environmental problems. And because of the close integration of tribal indigenous people into their ecosystems, the loss of biodiversity is closely tied to the loss of cultural diversity, the extinction of traditional tribal cultures.

This is especially important today. As the beliefs and practices of the industrial culture are being recognized as part of the global ecological crisis, there is an urgent need for a wider understanding of cultural patterns that are sustainable. The vast folk wisdom of American Indian, African, and Asian traditions has been viewed as inferior and backward by the industrial culture. It is time to reverse this Euro-centric arrogance and to recognize that many of these traditions - their ways of knowing, technologies, knowledge of foods and medicines, forms of aesthetic expression, patterns of social interaction, communal relationships, etc. - embody the ecological wisdom we so urgently need today./8

This is what I am saying since fifteen years, having founded, back in 1994, *Ayuda International Foundation* for the protection of tribal people's wisdom about life, and their high cultural diversity, and wistful traditions for healing and integration of emotions.[2] Yet it's a fact that in most developing countries technologies for recycling and for healing the badly afflicted nature are costly and not as accessible and readily available as in high-tech nations. Only truly supportive cultural and technological exchange between rich and poor countries can help changing this dim picture. Whatever our personal opinions in the face of these huge global problems we've got, and that also, and more sadly so, our next generations will be burdened with, we have to keep an open mind and learn, and change our rigid positions.

Fritjof Capra and Gunter Pauli have given in this reader very useful suggestions, and they can be taken as starting points for deeper study, as the field of investigation is huge, and never-ending. Nature's complexity is perhaps the single one most important topic of study for 21st century science, and I hope I can contribute a little to it by my own efforts. As for the authors of this book, they surely have done their very substantial contributions!

---

[2] ayudainternational.org

Published by Sirius-C Media Galaxy LLC, 2011

RIANE EISLER

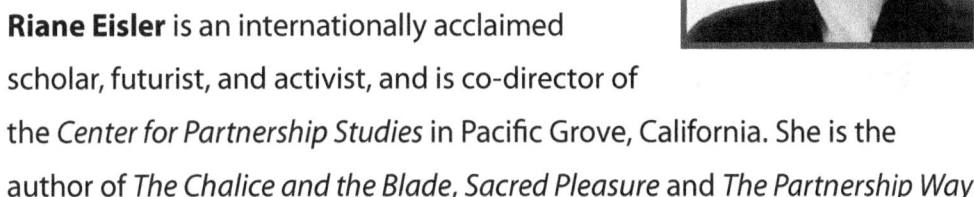

Books Reviewed

The Chalice and the Blade (1995)

Sacred Pleasure (1996)

Riane Eisler is an internationally acclaimed scholar, futurist, and activist, and is co-director of the *Center for Partnership Studies* in Pacific Grove, California. She is the author of *The Chalice and the Blade*, *Sacred Pleasure* and *The Partnership Way*.

I found Riane Eisler and her amazing research through a reference in one of the books by *Terence McKenna*. McKenna elucidated that some of the terminology he was using in his books and speeches was not entirely his own, and that, for example, the expression *dominator cultures* was one he had taken over from Riane Eisler, whose research he very much admired.

Having read most research about the old matriarchy-patriarchy dichotomy by *Johann Jakob Bachofen, Carl Gustav Jung,* other Jungian authors such as *Erich Neumann,* and first of all *Joseph Campbell,* I was wondering what Riane Eisler would have to tell me?

So I immediately bought her two major books *The Chalice and the Blade* and *Sacred Pleasure* - and started reading. And my eyes began to open wide …

RIANE EISLER

The Chalice and the Blade

Our History, Our Future
San Francisco: Harper & Row, 1995

Riane Eisler's research brought to daylight that we were stuck in some kind of neurotic scientism by upholding the age-old dichotomy of matriarchal-patriarchal when we describe evolutionary changes in the human setup, and that in reality we are dealing with a *partnership paradigm* versus a *dominator paradigm*, the first coming close to the idea of matriarchy, the latter more or less synonymous with patriarchy.

The merit of Eisler's approach is that we can get away from extreme positions: because there never was a really pure matriarchy or a really pure patriarchy in human history. When we look, for example, at the mythology of highly patriarchal tribes, such as the ancient Hebrews, we find matriarchal elements, and in highly matriarchal tribal cultures, such as the *Trobriand* natives in Papua New Guinea, we find patriarchal elements. This was found and explained at first by the Polish-American anthropologist *Bronislaw Malinowski*. Therefore must conclude that we got a mix rather than a pure soup, always. In that mix, to stay with the example of the Hebrews, there is a firm patriarchal root structure with a few matriarchal elements, as in yang is a small portion of yin. By the same token, and on the same lines of research, we can recognize in a highly matriarchal culture such as the Trobriand culture in Papua New-Guinea, patriarchal elements, as *Wilhelm Reich* found them, for this culture, in his research on matriarchy.

Published by Sirius-C Media Galaxy LLC, 2011

Already *Johann Jakob Bachofen*, one of the first authors on the matter, and who spent a lifetime with delivering abundant scientific research on matriarchy, found those patriarchal elements in all matriarchal cultures. Thus when we use the dichotomy matriarchal-patriarchal, we are arguing not from a real-life perspective, but rather from our ideological understanding of patriarchy or matriarchy, as if either of these were some kind of absolute values. But they are not. And to have come to this insight, I am indebted to Riane Eisler and her very well-researched books.

The Chalice and the Blade is a unique account, like a gigantic fairy tale, written by a woman who tells, perhaps with a crying eye, and another, angry, eye, all the violence, rape, and destruction that has been done in five thousand years of patriarchal madness! It's a unique book. And the author merits a Nobel prize or something like that for her research, and another medal for her courage!

This book, if you can read it cold-bloodedly, I would go as far as saying, you are probably from the same wood as those who are described in the book! This book drove me mad, and I got so infuriated at times that I had to stop reading, and continue another day, when I was in a more tranquil mood again. But Eisler, and here she's really heroic, always stays cool even in front of the greatest cruelties and atrocities she reports. She's not one of those feminists who turn the moralistic screw when it goes to write about child abuse. Eisler's books do not contain any polemics, which is why they are easy to read even for men, a fact that however does not mean that the information they contain is easy to digest! In a way, this book is a must-read for every educated human in our times. You just can't afford to not have read this book, so important it is. The applause that Eisler got from the scientific community is truly merited.

I will in no way paraphrase any of the lucid descriptions in this book, so let me comment on a few quotes that I hope the author and publisher permit me to put in this review.

Eisler actually suggests nothing less than a paradigm shift of reality, or she invites us to help her creating a new reality. She writes in the opening pages of the book:

Riane Eisler

But if we free ourselves from the prevailing models of reality, it is evident that there is another logical alternative: that there can be societies in which difference is not necessarily equated with inferiority or superiority./xvii

This is a very important point of departure, indeed. To tackle the old dichotomy patriarchy-matriarchy actually by overcoming it, is new, really new, in the whole of the scientific debate about it. It's new because most authors decided and decide pro one and con the other, thus resting within an ultimately invalid either-or scheme that was descriptive only, originally, as Eisler showed so lucidly, instead of being normative. But that it was descriptive and not normative, you had to read Bachofen for good, and not just secondary literature about Bachofen's life-long research.

The next important difference of Eisler's approach to all the other as it were conventional approaches to the matriarchy-patriarchy discussion is that she boldly shifts the observer point. Instead of talking about patrilinearity and matrilinearity, and thus about rather secondary rules of inheritance of property, Eisler looks from a relational point of view. She asks the question: 'How do males and females relate to each other in either of these models?' Well, and you clearly see the answer once you are able to lucidly formulate the question. The answer is that in the matriarchal setting, you got a sound

Published by Sirius-C Media Galaxy LLC, 2011

tenor about cooperation and participation, about complimentary elements in a mixed soup, that are of equal importance.

This is what Eisler then coined as the *partnership paradigm*. And after that you got the subordination of the female, you got different levels, and there was no more a mixed soup but two different soups, one that was cooked in the basement which was the female soup and one that was cooked on the 44th floor of the Tower of Babel, and that was the male soup. And from here the perspective of the male to be *on top* and *in control* and that of the female to be *below* and *out of control*. And sexually, we know that most natives copulate with bodies kneeled in front of each other, and thus in a basically equal position, while the standard sexual position in patriarchy, and our own culture as a result, undoubtedly is the missionary position, that has got its name for good reasons - reasons that we all know about, but to a lesser level want to talk about. Right?

This is what Eisler coined as the dominator paradigm, and the sexual 'positions' can serve as a metaphor for the social and even the anthropological positions. Native peoples were virtually fucked by missionaries, and often also in a less figurative sense, as we know it from countries like Brazil and others where militia and military have a well-defined and documented scheme of behavior when they raid native villages, and here the popular image of being fucked seems to reflect a sad reality. Eisler writes:

> **Riane Eisler**
>
> And if we look at our past – at the routine massacres by Huns, Romans, Vikings, and Assyrians or the cruel slaughters of the Christian Crusades and Inquisition – we see there was even more violence and injustice in the smaller, prescientific, preindustrial societies that came before us./xiv

Much of the rape that patriarchal history really is filled up with, and that let me coin the term *rape culture* for it, alongside with the term *murder culture*, is cultural in the sense that physical rape, the violent sexual penetration of the female by a *conquering* male, is only the smaller octave of a phenomenon that was from the start, when we only think of the Assyrians, was meant to be a military, social and political domination in the largest possible sense of the word.

With the Assyrians and many of the subsequent patriarchal tribes, the enemy was annihilated to a point that children and babies 'were smashed on the ground and against the walls', as the Bible reports, and whole cities were burnt down to ashes, after everything rapeable was raped and everything takeable was taken.

This was the typical scenario we find documented from all those early slaughters committed under patriarchal mankind - and here I justly and consciously use the older term mankind, and not the newer term humankind. For humans - they were not, excuse me, but that's my view of the matter. To be human, you have to give up your murder gods, your moralism, your draconian laws and your *righteousness*.

And sadly enough, in today's America, all these things are raising up again, and that at the same time the rape statistics, and the child rape statistics rise up in that *holy, free and pure nation* is really no wonder and only confirms my point of view. Or, as *Laurence G. Boldt* writes in his extraordinary book *The Tao of Abundance (1999)*, p. 55:

Laurence G. Boldt

There is no good that is always or only good, no single virtue appropriate to every situation. Lao Tzu tells us that it is only when we lose contact with out innate intuitive intelligence that we resort to goodness and righteousness as the ethical guideposts of our lives. It is only when real love is lost that we resort to filial piety or family values. It is in this spirit that we can understand St. Thomas Aquinas' dictum, Love God and do as you please. Love is superior to any ethical code.

Another important point mentioned by Eisler is the fact that our minds are pretty much conditioned by a single-cause etiology, as it were, of our cultural birth and origins.

Riane Eisler

In short, though only twenty-five years earlier archeologists were still talking of Sumer as the cradle of civilization (and though this is still the prevailing impression among the general public), we now know there was not one cradle of civilization but several, all of them dating back millennia earlier than was previously known – to the Neolithic./11

And more generally, her research has clarified that the barbarian primal horde that ghosts in the heads of so many scientists is just another of the myths we have been served with in school. Perhaps, yes, we became that barbarian horde under patriarchy, but that was *not* the source event, that was not the cradle of civilization, but a later stage of what I call devolution, and what Eisler herself calls the *truncation of civilization*. She lets no doubt that the pre-patriarchal civilizations, and among them first and foremost the Minoan Civilization, really were civilizations in the true sense of the word:

Riane Eisler

To say the people who worshipped the Goddess were deeply religious would be to understate, and largely miss, the point. For here there was no separation between the secular and the sacred. As religious historians point out, in prehistoric and, to a large extent, well into historic times, religion was life, and life was religion. One reason this point is obscured is that scholars have in the past routinely referred to the worship of the Goddess, not as a religion, but as a fertility cult, and to the Goddess as an earth mother. But though the fecundity of women and of the earth was, and still is, a requisite for species survival, this characterization is far too simplistic./ 23

And for all those who believe democracy was a 'modern' invention, Eisler explains:

Riane Eisler

Especially fascinating is how our modern belief that government should be representative of the interests of the people seems to have been foreshadowed in Minoan Crete long before the so-called birth of democracy in classical Greek times. Moreover, the emerging modern conceptualization of power as responsibility rather than domination likewise seems to be a reemergence of earlier views./38

It is in fact one of the strategies of the patriarchy-addicted detractors of truth about human history to veil the fact that not a single part of our 'modern' progressive worldview dates from any period after 1750. In fact, most of it dates from before 3000 B.C. And none of it is in any way new. There is nothing new in modern times, except the level of stupidity of the majority of human beings, that really was never as low as to-date. As Eisler wrote in *Sacred Pleasure*, her second book:

Riane Eisler

This is that one of the best-kept historical secrets is that practically all the material and social technologies fundamental to civilization were developed before the imposition of a dominator society./66

What Eisler of course did not mention in her fine analysis of patriarchy is the utter smeary hypocrisy that it is surrounded with, and imbedded in, and that in today's USA-dominated world media is coming along again with the pride of the Assyrian warriors of old, in a brazen wrapper that conditioned citizens take for brazen content.

Riane Eisler

Sacred Pleasure

Sex, Myth and the Politics of the Body
– new paths to power and love
San Francisco: Harper & Row, 1996

Riane Eisler's second book *Sacred Pleasure* is not less of a stroke of genius than her first, *The Chalice and the Blade*. In fact, both books are complementary in a way, they should be edited as a two-volume reader, in my opinion, from a publisher's point of view. The point of departure of this book is turning upside down most of our opinions about sexuality. I agree with Eisler when she says that most people are unaware of the fact that their sexuality has not just fallen from heaven but represents a carefully conditioned habit:

> **Riane Eisler**
>
> In short, sex does not, as a once-popular song had it, 'just come naturally'. Rather, as illustrated by the jarring differences in the prehistoric and contemporary sexual symbols and images we have been comparing, sex is to a very large degree socially constructed./22

As I have shown in my own writings, human sexuality is not, as modern sexology makes us believe, a matter of instincts, drives and automatisms. Sexuality could be entirely different from what we think it is, and what our sex laws make it to be, that is some kind of unwanted rip-off built in man by a laughing Gee Oh Dee who wanted to have some fun seeing tiny humans copulating their souls off in despair of the lost paradise. No, little man, you are not created as the little fucker in front of the Lord! You are not a *fucker* in the first place. Your sexuality is not a sex-machine, as you yourself are not a machine. But you don't know *what* it is, that is commonly called sex, and that's your problem. You don't see that it's energy, that it's God. Yes, sex is God. So, now, think about it. Why have you got to be sexual? And why have your children got to be sexual, if you want it or not? They got to be sexual because sex if life, because sex is energy, because sex is novelty, and creativity! I am especially grateful for Eisler's insistence upon describing Minoan Civilization as completely as possible in the context of her book. This really has great merit because very little is known today about this great civilization of pre-patriarchal times that was ruthlessly annihilated by the invading patriarchal hordes.

I will not further comment on this book but provide a few quotes to let the author speak for herself, for I feel not apt at paraphrasing her competent lecture. I think the book speaks for itself, while what the author says is all but self-evident. Sexuality has never been a comfortable issue in Western society and for that reason already, the book is not a comfortable read for everyone. But so much the better! It will shake some people out of their comfort zone ... to get them more connected with their bodies, and with life! This book merits not less praise than the *The Chalice and the Blade*, that I have equally reviewed.

Riane Eisler

The underlying problem is not men as a sex. The root of the problem lies in a social system in which the power of the blade is idealized – in which both men and women are taught to equate true masculinity with violence and dominance and to see men who do not conform to this ideal as 'too soft' or 'effeminate'./xviii

If we look at the whole span of our cultural evolution from the perspective of cultural transformation theory, we see that the roots of our present global crises go back to the fundamental shift in our pre-history that brought enormous changes not only in social structure but also in technology. This was the shift in emphasis from technologies that sustain and enhance life to the technologies symbolized by the Blade: technologies designed to destroy and dominate. This has been the technological emphasis, rather than technology per se, that today threatens all life in our globe./xx

In sharp contrast to later art, a theme notable for its absence from Neolithic art is imagery idealizing armed might, cruelty, and violence-based power. Nor are there any signs of 'heroic conquerors' dragging captives around in chains or other evidences of slavery./17

In Neolithic art, neither the Goddess nor her son-consort carry emblems we have learned to associate with might – spears, swords, or thunderbolts, the symbols of an earthly sovereign and/or deity who exacts obedience by killing and maiming. Even beyond this, the art of this period is strikingly devoid of the ruler-ruled, master-subject imagery so characteristic for dominator societies./18

Published by Sirius-C Media Galaxy LLC, 2011

MASARU EMOTO

Masaru Emoto is an internationally renowned Japanese researcher and an independent thinker. Certified as a *Doctor of Alternative Medicine* from the *Open International University*, he is also a graduate of the Yokohama Municipal University's department of humanities and sciences, with an emphasis on *International Relations*. Masaru Emoto's research has visually captured the structure of water at the moment of freezing, and  through high-speed photography he has shown the direct consequences of destructive thoughts and the thoughts of love and appreciation on the formation of water crystals. The revelation that our thoughts can influence water has profound implications for our health and the well-being of our planet. Masaru Emoto has written many books, including the New York Times bestselling *The Hidden Messages in Water* and *The True Power of Water*. - From: *The Secret Life of Water*

Masaru Emoto

The Hidden Messages in Water

New York: Atria Books, 2004

I became aware of Masaru Emoto's water research through the film "What the Bleep Do We Know!?"[3] It was an information that really left me speechless, and I ordered his books at once, and got all the information I could get from the Internet about him and his amazing research. I will not enter in my book reviews the discussion about credibility of his research or his degree (from *Life University*, Calcutta) that some of his detractors focus upon to tear down his reputation. Allegedly, as the Wikipedia article[4] points out, his water photography-technique is not meeting the standard of double-blind tests. So I leave all this open and will base my book reviews strictly on quotes taken from the books, and try to check his research back with my own twenty years of bioenergy research.

Surely, Masaru Emoto's research has hit the rock, so to speak, it has moved the earth, it has made huge waves, it has mobilized funding, it has got people to change their entire life, their worldview, their inner setup – and for good reason. What I can't understand after all is why we in the West did not catch this information much earlier? I have researched now for fifteen years

---

[3] See my preview of this famous movie:
http://blip.tv/reviews-and-previews/what-the-bleep-down-the-rabbit-hole-by-betsy-arntz-preview-by-pierre-f-walter-2302832

[4] http://en.wikipedia.org/wiki/Masaru_Emoto

Published by Sirius-C Media Galaxy LLC, 2011

on the bioenergy, I have read the major esoteric writings of the world, studied the *Mystery Schools*, and the ancient hermetic tradition is nothing new to me as I studied it already twenty years ago, but why did I never come across the concept of *hado?* Why did I have to learn that from Dr. Emoto?

I will now try to bring some structure in this book review by relying not on hearsay, so much the more as Dr. Emoto remains to be controversial as a scientist, but on the author's own statements. I will put some quotes and comment on them. If Emoto has not got full credentials, so let us look what his writings reveal about him. I got the impression that he comes over as a wise man, a man with a very high level of intuitive knowledge, and also somebody who knows to write. His style is easy yet deep. He knows to express truth in a very clear and pristine way. To begin with, he writes:

> **Masaru Emoto**
>
> So how can people live happy and healthy lives? The answer is to purify the water that makes up 70 percent of your body./xvi

If that is not a clear statement, I don't know what is one. Frankly I have never considered before in my life the fact that I consist mainly of water, and that because of this simple fact, I have to do something about that water I am consisting of. Have you? To be true, only Paracelsus, one of the greatest healers in human history, and whom I have studied at length, reading his writings early in my life, in their German original, said something similar. And my next question, logically so, would be: and why *water?* Emoto replies:

> **Masaru Emoto**
>
> Water serves as a transporter of energy throughout our body./xvii

Having studied virtually all written traditional knowledge about the bio-energy, after so many years, I overlooked the most essential and thus had to learn it from Emoto, that is that this *ch'i* that flows through my body flows through my body because of water, because it basically flows through that watery substance in me. Now, Emoto, puts it more precisely:

Masaru Emoto

More now than in the past, the medical community has begun to see water as a transporter of energy, and it is even being used in the treatment of illness. Homeopathy is one such field where the value of water is recognized./xvii

Homeopathy is indeed concerned with water. But we hardly ever knew why? We hardly ever knew why a homeopathic formula is diluted so much, and consists almost entirely of water? When we get to know that water is the magic here, and not the substances that are mixed with it in a homeo-pathic tincture, all becomes clear. Succinctly speaking, there are two major arguments that Emoto advances in order to explain his research, and that his detractors do not seem to catch up with. What is it that makes water to be a receptor and vehicle for thought? I think it is the fact that water, as all in life, is *vibration* and that his vibration can be manipulated through intent. Emoto writes:

Masaru Emoto

What you really know is possible in your heart is possible. We make it possible by our will. What we imagine in our minds becomes our world./xxii

Now, how does the alteration of vibration come about? Emoto explains:

Masaru Emoto

The lesson what we can learn from this experiment has to do with the power of words. The vibration of good words has a posi-

Published by Sirius-C Media Galaxy LLC, 2011

tive effect on our world, whereas the vibration from negative words has the power to destroy./xxv

Now, in fact this is true. The hermetic tradition taught since times immemorial that words are codified vibrations. The scriptures all converge in saying that in the beginning there was the Word, and that the Word was sacred and had creational power. In old Egypt and India, as Manly P. Hall writes in *The Secret Teachings of All Ages (1928/2003),* p. 2, the hierophants used vibrations for healing:

Manly P. Hall

The magic rituals used by the Egyptian priests for the curing of disease were based upon a highly developed comprehension of the complex workings of the human mind and its reactions upon the physical constitution. The Egyptian and Brahmin worlds undoubtedly understood the fundamental principle of vibrotherapeutics.

More generally, Jonathan Goldman, a present-day vibrational healer, writes in his book *Healing Sounds (2002),* p. viii:

Jonathan Goldman

Everything is in a state of vibration. Everything is frequency. Sound can change molecular structure. It can create form. We realize the potential of sonic energy; we understand that virtually anything can be accomplished through vibration. Then, the miraculous seems possible.

Now, there is one more catch to understanding the background of Emoto's expertise, that is the Japanese *Shinto* tradition. Emoto writes:

Masaru Emoto

In Japan, it is said that words of the soul reside in a spirit called kotodama or the spirit of words, and the act of speaking words has the power to change the world./xxvi

Regarding human beings, the fact that we vibrate, that we are a bunch of frequencies, has been affirmed by not only the hermetic tradition, but also by clairvoyants. Not only do we vibrate, but we vibrate *differently*. In a sense, we all come with a unique vibrational pattern. For example, Shafica Karagulla writes in her book *The Chakras (1989)*, p. 2:

Shafica Karagulla

It is said by some that every human being emits a unique tonal pattern which is created by his individual energy fields working in unison. This is sometimes referred to as the personality note.

Emoto confirms this to be true equally from the perspective of the Shinto tradition and esoteric Japanese knowledge about the bioenergy:

Masaru Emoto

Human beings are also vibrating, and each individual vibrates at a unique frequency. Each one of us has the sensory skills necessary to feel the vibration of others./41

In his second book, *The Secret Life of Water*, Dr. Emoto has given more information about the specific vibration of water, which is knowledge seemingly only existing in Japan. I will thus discuss this point further in my review of *The Secret Life of Water*.

As far as the present book is concerned, I would like to add one interesting detail that was a surprising result of the experiments with exposing water to positive affirmations, negative affirmations (insults) or else leaving the water completely unattended. The surprising outcome was that the worst water, the one with the worst crystals, was not the water that had received the insults from the school children who helped carry out the experiment, but the water that had received no attention at all from their part. Emoto comments:

Masaru Emoto

To give your positive or negative attention to something is a way of giving energy. The most damaging form of behavior is withholding your attention./65

This is a fact known from research on *child abuse*. Children who have been abused tend to go back to their abusers despite the fact that abuse is going to continue. And there was always a question mark in forensic research why children do that, and why they do not, or very seldom, betray their abuser in order to get rid of the abusive relationship? It has been found that it's because the negative attention children receive in the form of abuse is for them still better than the total lack of attention they get in their homes. And this motivates us to perhaps to render our education more attentive to the true needs of children, as Krishnamurti emphasizes it in his book *Education and the Significance of Life*, because attention and love are one and the same thing. Try to show somebody that you love him or her and try to do that without giving them any attention. You will see that it's impossible. The very thought of the person is already attention, and by thinking of the person you are sending out a vibration, and energy.

To summarize this review, and to focus on the uniqueness of Emoto as an author, I would like to stress the simplicity and the clarity of his language, and his style. There is light in it, do what you will, say what you want. There are no convolutions in his style, there are no open questions. All is clear, pristine, like pure water. And here his message, and his own style, and probably all of his life, beautifully converge. This is how it should. And it proves that this man walks his talk. I have been enchanted by all his books.

Masaru Emoto

The Secret Life of Water

New York: Atria Books, 2005

The *Secret Life of Water*, when you compare it with Emoto's first book, *The Hidden Messages in Water*, is something like the scientific back office of water research.

Here, Emoto really explains what *hado* is, this strange concept that seemingly was unknown in the West, except among natural healers and clairvoyants. And yet it is a very old concept, part of the Shinto treasure of ancient Japanese wisdom, and thereby part of perennial science. Once I got familiar with this knowledge tradition, I found a number of other books about *hado*, as for example sending out *hado* by deliberate intent for healing, or learning the *hado of cooking*. Myself a passionate cook since forty years, I always wondered how it is possible that two people using the same recipe, and the same kitchen for cooking the same food can end up with cooking food that tastes differently. While the dish may even look the same, the taste is different. The mystical nothing that the Western mind explains away as *illusion*, the Japanese put in very precise terms, saying that the cook whose dish tastes better has a better or more sublime *hado*. I have even found books how to deliberately improve your cooking *hado* so as to cook better-tasting food, while you may cook the same food that you always cooked before. If this is not something Westerners will be intrigued about, I don't know what it can possibly be that will rock your life? Now, let's go step by step and inquire further, along with

Published by Sirius-C Media Galaxy LLC, 2011

some quotes from this very well-written book. Emoto enumerates three basic keys for the understanding of hado:

Masaru Emoto

Three key words are helpful to understand hado. The first is frequency. The entire universe is vibrating at a particular and unique frequency. Frequency can be modeled as waves, a fact easily supported by quantum mechanics. All matter is frequency as well as particles. What this means is that rather than considering something a living organism or a mineral, something we can touch or something we can see, everything is vibrating, and vibrating at a unique and individual frequency./30

The second word that is helpful in understanding hado is *resonance*. Resonance comes in play when there is a sender of hado information and a receiver of the information. Say you make a call to someone you want to talk to. Unless that person picks up the receiver, there will be no conversation. Without a receiver, information cannot be sent. The Japanese expression *aun no kokyu*, or 'in-breath and out-breath', describes a state where subtle synchronization occurs when we do things together. This also refers to a relationship between a sender and a receiver. When there are vibrations matching, resonance occurs. We can observe the phenomenon of resonance in various aspects of daily life. For example, if you have feelings of hatred toward someone, there is a good chance that this person feels the same about you. Likewise, if you have positive feelings toward someone, that person will sense those feelings even if you don't express them in words. What we feel in our hearts has a strange way of being relayed to other people./32

The third word helpful for understanding hado is similarity. The macro world we know is a symbol, an expansion of the micro world. The nine rotating planets in our solar system are the macro version of the electrons circulating around the atomic

nucleus, and what is going on within the human body is a minia-
turization of what is going on in the grandeur of nature./33

This is not all about hado, it's just a starting point. From about page 50 of the book, Emoto expands about healing with hado. And he has collected amazing examples from all over the world, and from different researchers, to prove his point. He envisions what he calls *hado medicine* becoming one day the medicine of the future. He writes:

Masaru Emoto

All symptoms of illnesses vibrate at a unique frequency. By knowing the frequency, it is possible to overlap the exact opposite wavelength on top of the symptom's wavelength; thus, the frequency of the illness is dissipated and the symptoms are alleviated./51

This is indeed something that has been done by the Russian-French researcher Georges Lakhovsky who, as early as in the 1920s, was able to heal plant cancer simply by exposing cancer-afflicted plants to vibrations that were exactly opposite to the frequency of the malignant cells. From these experiments, Lakhovsky then elaborated a cancer etiology and sound healing procedures for both plant cancer and cancer in animals and humans. For Emoto, the body is something like a complex sound machine and it really vibrates, emits frequencies and can be seen as a musical composition. All organs produce sounds, and all the sounds are in harmony with each other in the healthy organism. Now what happens when we are sick? Emoto explains:

Masaru Emoto

When something goes wrong somewhere in the body, there is discord with one of the sounds. And when even one sound is out of pitch, the entire composition is not as it should be./52

Published by Sirius-C Media Galaxy LLC, 2011

A controversial point in Emoto's science of hado is what he calls the *memory of water*. He claims that all water has a memory that manifests through the fact once an affirmation has been emitted, and water has been impregnated with such positive or negative intent, this impression lasts. It will not just vanish after a day or a month. But how can we imagine this in practice, and what are the details of this science? How long will the impression last in the individual case, and how to detect it? This seems to be a floating science, for it appears to lack specific data, if I am not mistaken. Emoto expresses himself in terms that can neither be criticized, nor taken as evidence for the memory theory:

> **Masaru Emoto**
>
> All matter has its own hado, and water relays this information. The molecules of water carry messages like the magnet of a computer disk. Hado can be either beneficial for life or harmful for life. But even if the vibration is good for life, if water - the mediator - is impure, the hado will not be relayed correctly./62

As I mentioned already in my review of *The Hidden Messages in Water*, Emoto's research is controversial with regard to scientific standards applied. While he seems to have given contradicting information to the press in this regard, in the present book he writes, quite honestly:

> **Masaru Emoto**
>
> I admit that the selection process is not strictly in accordance with the scientific method, but simply put, we choose the crystal that best represents the entire sample instead of simply one from the most common category./130

The fact is namely that there is never a total uniformity in the water crystals that are formed after the water was impressed, and impregnated, with intent. There is always a mix. Now, when there is a mix, which crystals are going to be photographed and shown in a publication? It's well clear that this is a crucial point in the whole of this research. To argue from the detractor position: if there is a mix, there is no proof at all because when there is a mix, all is *potentially* in there, and so I can just pick out what I like to pick out, and comment on it.

Now, strangely enough, Emoto doesn't even come up with the idea of a *predominant* scheme of crystals so that we could establish something like a rule of evidence based upon majority of crystals versus minority of crystals. The fact is that Emoto not only applies intent for choosing the crystals but he also applies intent for choosing the choosers. He has argued in interviews that he was carefully selecting the people who were doing the photographs because another crucial point brought forward by the detractors was that if intent is so powerful on water, then what about the intent brought in the water, more or less unconsciously, by the photographer?

And how can we detect to what extent the crystals have been formed by the affirmations, glued as paper messages on the bottles, on one hand, and the intent fostered in the minds of the photographers, on the other? I think I can dare to carefully put a question mark here as to scientific credibility. While I intuitively agree with Emoto and his research, I think its scientific foundation is far from being established.

Let me leave a note on binding. *The Hidden Messages in Water* is a not-so-lucky paperback and already after the first lecture began to unbind from the back cover backwards. I really hate paperbacks that unfold because the glue has dried out or was not properly applied. Not all paperbacks do it, but generally there is nothing more beautiful for publishing a book than a hardcover. And the present volume is a very well-done hardcover, and I wonder why it's almost double the price? I can produce a hardcover edition for only 10 to 20 percent more than a paperback, and in some printing companies even for about the same price.

Published by Sirius-C Media Galaxy LLC, 2011

To charge for a hardcover practically twice the price of a softcover is unreasonable, and it shows that the publisher has no control over the labor-intensive part of his business. Book-binding is not machine work, it's human work, and that adds on to the beauty of it.

And I hope that this message will get through to American publishers. You can fabricate a hardcover in India for less the price you produce a softcover in the USA. I am sure most people would put a few dollars more for a hardcover if publishing multinationals were more reasonable on their added charges for them. And if they cannot outsource their production to cut costs, why are they calling themselves *multinationals* in the first place?

I know that this is not primarily an American problem, as for example in Germany, production costs for hardcover editions are even higher, while almost all the printing machines in the world are from Germany. But myself being German, I am aware that Germany today has lost almost all its competitiveness and cutting edge in business except for its tight range of machinery that it can admittedly produce and sell very well. But generally, production costs are much too high. For the USA, with its huge creative resources, I am sure the hardcover book sales could rise up significantly with applying outsourcing and production abroad. The customer would value the much higher durability of a hardcover, for a few dollars more, but surely not for double the price.

FENG SHUI BOOK REVIEWS

7 Feng Shui Books Reviewed

Dr. Ong Hean-Tatt, PhD

Amazing Scientific Basis of Feng Shui

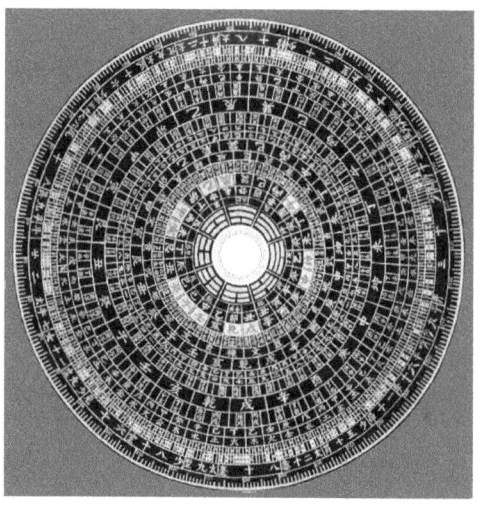

Evelyn Lip

The Design and Feng Shui of Logos

Karen Kingston

Creating Sacred Space with Feng Shui

Lillian Too

Feng Shui

Man-Ho Kwok

The Feng Shui Kit

Nancilee Wydra

Feng Shui, The Book of Cures

Richard Craze

Feng Shui - Feng Shui Book and Card Pack

Feng Shui, now booming as the first and foremost technique of living in harmony with cosmic energies, is by its very nature alien to Western thought. This is so because in the West, the perception of life as an energetic process that is highly complex yet also highly flexible became taboo with the dominance of powers that tried to control life and man. Knowledge about life was forbidden and left to alchemists who risked their lives pursuing the only real science that existed at that time. Even with the so-called scientific era of humanity, the knowledge taboo persisted and the dominance only changed its camps. What formerly was the privilege of Church officials, was then, and is until today, in the hands of scientists.

In Asia, however, this knowledge-prohibition never existed and *Feng Shui*, the Chinese science of the cosmic life energy and its manipulation for health, power and happiness, is not only a set discipline for every scholar, but has its roots in popular tradition.

These book reviews show both the Eastern and the Western approach to the perennial science of *Feng Shui*, and it becomes obvious that this science is something that fascinates high-caliber scientists such as Dr. Ong from Kuala Lumpur as well as housewives who may pick a Feng Shui card to assess if the oven is best put in the corner or in the middle of the kitchen …

This, then, is again typical for Asian scientific concepts; they are never pure theory but have their practical stance in life, and if not, they won't be pursued at all. The problem is that Western thought is so much pervaded with the dichotomy science versus ordinary life that people in the West tend to judge these practical applications of Feng Shui in a depreciative way, saying that Feng Shui was 'the art of furniture arrangement' and other reductionist statements of this kind. When the science of Feng Shui explains how the vital energies move in our house, and accordingly advises to arrange furniture in a certain way, and not in a certain other way, then this is not a reason to deny this science its very status as a science, and relegate it to the areas design, lifestyle, fashion, favorite interests, and all the rest of the modern consumerist soup.

While it was admittedly designers who first discovered Feng Shui in the West, this doesn't mean that Feng Shui is *but* a design tool, as many people tend to affirm. That would be like saying that electricity is not a science be-cause a light bulb is 'after all a piece of furniture'. But it is exactly this kind of reductionist thinking that has precluded the understanding of Asian scientific concepts for so long.

Published by Sirius-C Media Galaxy LLC, 2011

Dr. Ong Hean-Tatt

Amazing Scientific Basis of Feng Shui

Kuala Lumpur: Eastern Dragon Press, 1997
(No cover scan available)

Amazing Scientific Basis of Feng Shui is a treatise - nothing less. It is not your usual Feng Shui book. The analysis of Dr. Ong reveals that the science of Feng Shui is an integral part of a perennial science concept about the cosmic life energy that can be traced back, in Oriental cultures, until Antiquity. The strength of the book is the unusual and wide regard of the author on a matter that, in the trend of New Age enthusiasm, is sometimes treated in a rather esoteric and unscientific manner.

Dr. Ong's approach is deep, inter-disciplinary and synthetic. Furthermore, the book is rich in resources. The author must have done extensive research across various sciences to have reached at his amazing conclusions.

Before the Dr. Ong starts his cross-science survey, he carefully prepares the reader and introduces the concept of Feng Shui as a science about the life energy that can be compared to acupuncture and that possesses similar knowledge about the subtle energy that is at the basis of the now recognized medical treatment of acupuncture. Both concepts, Dr. Ong shows, are based upon the same point of departure: the existence of a polarized bio-energy that the Chinese call *ch'i*, the negative part of it being called *sha*.

Then, the author examines and explains the truth expressed in ancient myths and legends like Homer's Iliad and advances evidence that such stories are to be taken literal and not just as metaphorical, occult or esoteric.

The book proceeds logically, starting from the science of Feng Shui and its historical development in China, then extending to the broader concept of *geomancy* that Feng Shui is historically part of and that has also a long tra-

dition in the West. Yet unlike typical Feng Shui books, the author does not enter the subject in practice, but focuses primarily on collecting and demonstrating scientific evidence, mostly from Western sources, in order to corroborate the view that Feng Shui is a *science* - and not superstition or magic, nor a design tool or fashionable lifestyle concept.

The regard that Dr. Ong takes upon the bioenergy is most interesting as he examines the traditional Chinese Ganzhi system, comparing it with the *Cabbalistic Sefirot* system, and then shows that Feng Shui can be compared with these sources of perennial wisdom and expresses the same truth in other words.

After an in-depth analysis of the I Ching, the *Pa Kua* (I Ching Based Feng Shui Compass) and the *Lo Pan* (Geomancy Based Feng Shui Compass), the author devotes three chapters to the examination of the concepts of ch'i, sha and the five elements - an invaluable source of knowledge especially for the Western reader and scientist. As a matter of fact, Dr. Ong shows the amazing similarity of the Chinese Science of the Five Elements with what he makes out as the Western Five Elements Concept. From there, he proceeds to examining the Western geomancy art of dowsing and provides evidence for the fact that there is a broad consensus between Chinese and Western sources regarding the detrimental effects of underground water and cancer-producing fault lines.

The factual evidence that Dr. Ong cites is of great value so much the more as these sources of evidence are difficult to make out.

But the author does not limit his research here. The way Dr. Ong explains the UFO phenomenon is among the most interesting and original approaches I have heard of until this day. Many of us may have doubted not the phenomenon as such, but the explanation for possible presence of extra-terrestrial forces on earth. Actually, paranormals and scientists involved with parapsychology have since long questioned this theory and put forward their view that those appearances rather are concerned with earth-bound inhabitants of a different, generally higher, energetic vibration than humans, that are invisible and live in part in the mountains and in part in the deep sea, particularly in the Bermuda Triangle.

Dr. Ong's theory is different: he assumes that UFO-related phenomena are brought about by spirits or ghosts, ectoplasms or thought forms created by human beings. To corroborate his theory he refers to the research undertaken about the Poltergeist phenomenon that showed that Poltergeists are emanations from the mind of highly electric nervous adolescents. This evidence then, undertaken in the 1970's at Stanford University on the young Uri Geller, cannot seriously be contradicted.

The logical parallel that Dr. Ong forwards here is highly original, and it is highly probable. It is perhaps more probable than not that those phenomena are actually part of our own sphere while they may be part of other life dimensions or vibrational fields that we do not fully understand yet. Other evidence that corroborates Dr. Ong's view of the UFO problem is the UFO grid, an astounding phenomenon that relates to the universal *Feng Shui*

Dodecahedron World Grid the existence of which Dr. Ong proves with convincing arguments and factual backup.

This universal bioenergetic world grid has its roots in ancient times and was known in Greek philosophy as the *12 Pieces of Skin* or in Russian esoteric philosophy as the *Dodecahedron Crystal*. The factual evidence produced by the author that relates in detail to various UFO sightings and reports from reputed sources is dumbfounding and seems to prove the fact that all these phenomena feed upon earth energies or telluric energies that are emanating from underground water, *ley lines* or monuments that have perhaps been erected for this purpose by the giants or angels, such as the pyramids, Stonehenge and others. The amazing fact is namely that all these monuments can be shown to have been built exactly on the crossing of two intersecting ley lines, which are telluric energy paths in the earth's aura.

Furthermore, Dr. Ong examines the bird migration phenomenon and finds that it corroborates the evidence forwarded for the existence of the world grid - the fact is that the birds more or less follow those lines and that the energy that emanates from them serves the birds as a navigation help.

The author then refers back to Oriental sources of knowledge and examines the Middle East concept of dragon energy forces and the basics of the Chinese *Water Dragon Classic* in order to prepare the reader for a still larger perspective: the links that exist between the global knowledge of telluric lines, the Feng Shui concept of fault lines, the link between Feng Shui lines and megaliths.

Further, the author makes Feng Shui to be understood from a perspective of the Western scientist who applies his own known concepts to the ones that are not yet officially integrated into Western science. For this purpose, Dr. Ong puts forward evidence that Feng Shui actually establishes a more

Published by Sirius-C Media Galaxy LLC, 2011

unified and harmonic form of perception that cares for a balance between the right and the left brain hemispheres.

At the end of his extensive study, Dr. Ong further deepens his UFO theory and asks if we are not dealing here with living beings instead of machines? He shows the links that exist between ley lines and the appearance of those spirit-UFO's which can be said to represent an extraordinary evolution of the whole of UFO research.

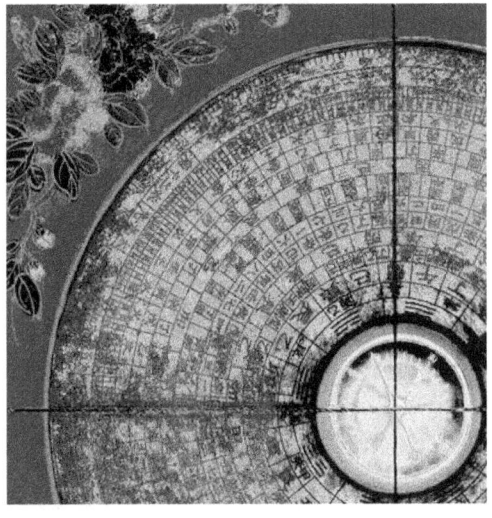

The study then proceeds to backup the findings with other evidence from such different sources as the March of the Lemmings, the 12-years *Sun Spot Cycle* and the deeper esoteric meaning of the word Feng Shui and the other Chinese words ch'i and sha.

My impression about this book is very positive. The reader of this review may have got the feeling that Dr. Ong's study is too vast and therefore difficult to understand or to read. The contrary is the case! Since years I have not got a book in my hands that I devoured with similar amazement, curiosity and inner tension. It was actually more like reading a fascinating novel than a dry scientific study. This is to say that Dr. Ong knows to write which adds to his amazing synthetic thinking capacity. The book has enriched me and given me encouragement to proceed with my own research on the still little understood functionality of the cosmic life energy. Many of the sources cited and explained by Dr. Ong actually corroborate my own findings about the cosmic energy.

There is only one negative point to mention about this book. It was obviously difficult for Dr. Ong to find a publisher in his home country Malaysia that provides him with a correct editorial work and spell-checking. The book definitely does not comply with Western or international standards of editorial work.

Evelyn Lip

The Design of Feng Shui Logos, Trademarks & Signboards

Singapore: Prentice Hall, 1995

The Design & Feng Shui of Logos, Trademarks and Signboards
by Evelyn Lip is a really useful booklet about the successful design of company logos and design features used in stationary or advertisement. As the author rightly remarks in the Preface of the book, there are indeed very few books on this important subject. In practice, many company owners or even large corporations leave it completely in the hands of designers to brand their company image. There are of course good reasons for doing this, for not everyone has got aesthetic sense or the capacity to draw a logo free-hand or on the computer.

On the other hand, nobody else than yourself, founder, owner, CEO or General Manager of an organization can feel what you want to convey to the general public with the image you are going to present as a mark and symbol for what you are doing or intending to do.

Designers are now slowly catching up and some of them have implemented principles of Feng Shui in their designs. But those designers are still the exception, at least in the West. There is not only the aesthetic side in design. Feng Shui namely teaches us that in every design or symbol there is also an energetic impact. Actually, Carl Jung has found the same to be true in his research on symbols and archetypes. Into every symbol an energy is woven which can be used either positively or negatively, depending on the intention of the one who uses it, but depending also on his or her knowledge about the power of symbols.

Symbols are often archetypal in character and thus express content that is common to all of us, through our being connected within the universal spirit network of the collective unconscious. Feng Shui expresses this in other terms but it means the same. In the Introduction, Evelyn Lip helps raising our awareness regarding the importance of impacting on our business destiny through taking appropriate action.

The booklet very carefully examines and discusses successful and less successful logos and signboards after a brief introduction into the cultural impact of logo design. Among the design principles that the author outlines and marks as more or less general Feng Shui design principles, is the principle of *less is more*. In her discussion of this principle, the author discusses the *Lufthansa* logo as an example, saying that the emblem is graphically simple yet significant and is perceived as synonymous with quality, forming the basis for a sophisticated information system.

Another principle that the author qualifies as a Feng Shui principle for logo design is the expression of unity, strength and harmony. Here, as an example, the author depicts the Chase Manhattan Bank logo for further illustration.

The booklet is a must in the library of every logo design company and a useful item to study for every business owner. Usually what happens while reading the booklet - this is my own reading experience - is that from the quantity of logos depicted and discussed, a kind of vague feeling comes up for the logo that one searches for one's own business. In my case, the feeling I got was interestingly to abandon all my previous ideas of a logo that uses a picture related in some way to my business, and instead use a very simple text logo without any graphic.

Published by Sirius-C Media Galaxy LLC, 2011

A pity only that such a lucid book has got such a mediocre cover design! I have flattened the awful color setup and the even more dreadful jpg-compression used by amazon.com, to reproduce the cover here in simple b/w. It's beyond my grasp how an author who writes a book about design can let all of this happen! It's perhaps just another indication of many I have collected over the years of how little support and empowerment authors get from the publishing industry.

Karen Kingston

Creating Sacred Space with Feng Shui

New York: Broadway Books, 1997

Creating Sacred Space With Feng Shui is much more than a Feng Shui book. It is a whole life philosophy, and written from a strongly personal yet nonetheless verifiable perspective. This book is not a boring manual of Feng Shui. It is a most original piece of writing, unheard-of in some way.

Karen Kingston's unique talent is an inborn and absolutely stunning natural sense for the higher dimensions of existence, for all that is invisible to our physical eyes and undetectable for our five senses. The novice reader may be astonished about the authority that this text reveals and the power of the author's approach to Feng Shui that is the pragmatic and direct approach of an experienced practitioner.

There is something almost magic about this book. I have been immersed in it twice, this book being one of the very few books I have read more than once. The charm and the intuitive wisdom that the author transmits in an invisible way has kept me fascinated until the last word - and this equally during my second study.

But this book is not simply one of those poetic writings that elaborate a magic view of life. It is that also, but it is much more than that. Behind the beautiful appearance and the refined language is hidden a hard core manual that is truly scientific - in the sense of a higher and holistic form of science. Of course, the representatives of today's reductionist modern science would question most of Karen Kingston's scientific concepts. But this argument is true for almost all publications about Feng Shui.

What Karen Kingston does is exactly to go beyond the limits of a Cartesian science that is based upon wrong premises about life. To call Karen Kingston's approach to life or spirituality animistic, an argument that has been put forth even against such enlightened spirits as Johann Wolfgang von Goethe, Emanuel Swedenborg or Carl von Reichenbach would totally disregard the deep and intuitive truth that is at the basis of this holistic life philosophy.

Karen Kingston who is married with a Balinese and who lives several months every year in Bali, gives pertinent information in her book about the ways that Balinese use Feng Shui or Space Clearing. In my own experience, there are in Bali actually two levels of handling spiritual wisdom, a professional level – if I may say so – and a popular or intuitive level. The professional level is since many generations in the hands of the first caste, and especially the Pedandas (Hindu priests).

Here, we encounter a highly sophisticated and informed way of handling spiritual information that is so complex and so deep that most Westerners usually only shake their heads when they first hear about it. On the other hand there are 'the people in the street' who, in Bali, it seems, are also wiser as anywhere else in the world. For they, too, have this knowledge, only in a more intuitive and less literary form.

Having lived and worked myself in Bali for several years, I understand Karen Kingston's natural affinity with Bali and the Balinese. I could not imagine where else somebody like her could live. Following the book's advice and information about Balinese temple bells that are wonderful for clearing space with sound, I have myself acquired a Balinese temple bell, space-cleared my villa in Bali with it and can fully confirm Karen Kingston's detailed description of all the benefits that the sound of these bells has on the whole of our organism.

It all sounds like a miracle but I am convinced that it is all but magic and we will fully understand it once we know more about resonance phenomena, the complex influences that sound and vibrations have on our aura, on all our etheric bodies. The ancient musical healers knew all about it as the myth of Orpheus reveals. Compared to an approach to Feng Shui such as the one of Lillian Too, Karen Kingston seems to go much too far in her definition of Feng Shui. But it would be reductionist to trace such kind of borders.

Anyway, there is no doubt that Karen Kingston's approach to Feng Shui is one of the most original ones existing presently on earth.

Published by Sirius-C Media Galaxy LLC, 2011

Lillian Too

Feng Shui

Kuala Lumpur: Konsep Books, 1994 (Book)
Kuala Lumpur: Konsep Books, 2003 (CD)

Lillian Too is well-known in Asia. Her books are to be found in almost every major bookstore. The old science of *Feng Shui* has great appeal for the Asian public. Lillian Too is best suited to engage in this science. She used to be a very successful business woman who was, before she settled to write book after book about Feng Shui, President & CEO of a large bank in Kuala Lumpur, Malaysia. Her Chinese family tradition, too, may have contributed in her interest in this traditional Chinese science. But her success is certainly also due to the fact that business people in Asia have a high regard for Feng Shui, which is for the most part ignored by Western business people, even those living and working in Asia.

In fact, it has taken generations until the West awoke from its materialistic trance and inquired if there was not something more subtle in life than what the five senses can grasp. And now, paradoxically, it is for the most part the Oriental approach to geomancy, Feng Shui, that is becoming popular, and not our own Western geomancy tradition which is as erudite as its Oriental counterpart.

Dr. Ong, in the book that I discussed earlier in this book review, reminds us that the Druid sages were reported to be able to ride on the subtle energies so that they could fly in the air without any device other than the forces of the sun and the moon that they knew to activate for their purposes. Of course, the modern reader is less interested in these stories than in receiving practical and down-to-earth advice how he can improve luck, health, happiness and wealth in their lives. And this pragmatic kind of approach is exactly the tenor of most Feng Shui books that are published in Asia. Lillian Too gives such advice and she does it in a straightforward manner that is exemplary. Her working together with renowned Feng Shui master Yap Cheng Hai led to a fruitful collaboration that the books brilliantly testify.

Lillian Too defines Feng Shui as *The Art of Living in Harmony with the Land*, the deeper wisdom that teaches us the power of being in harmony with all our surroundings. A wisdom, to repeat it, that was taught equally in the West, all through our history, that however was disregarded by the power institutions like the Church and later the nation states that were not interested in the subtle truth of life.

The book proceeds teaching Feng Shui in twenty-three chapters. The style of the book is very methodic and it presents knowledge in the traditional, deductive way. Whereas now in the West the trend goes more in the direction of presenting new and unusual knowledge inductively or empirically. This rather traditional and academic way of presenting content may at times contribute to a somewhat boring reading experience. However, the author knows to act counter to this danger by giving many examples of her practice, telling anecdotes or referring to the writings of the old masters that she appears to have studied extensively.

The book is certainly a very good starting point for more in-depth Feng Shui studies and practice, and so much the more for readers who are trained as scientists or in any other traditional, methodic way.

Published by Sirius-C Media Galaxy LLC, 2011

Man-Ho Kwok

The Feng Shui Kit

London: Piatkus, 1995

The Feng Shui Kit was the very first book I found on the subject of Feng Shui, ten years ago. And it is more than a book! On the back of the cover I had written *Bookstore Hyatt Regency*, Surabaya, Indonesia, April 3, 1996.

This Kit had a very strong impact on me, not only because it had been the first publication that got in my hands about the fascinating perennial science of Feng Shui, but because it was an item to play with, something I could become active with right away. For there is a *Lo Pan* delivered in the kit, which is a Feng Shui Compass, though from plastic. But it contains the basic signs and directions, and in the book you can look up the explanation.

To give a practical example. The first thing to do is to adjust a wheel on the compass the way that the red-white compass needle is exactly in the middle of two little red points. The direction here indicated is South. The compass actually possesses two wheels that you can turn. The greater wheel that is behind contains your personal animals in Chinese astrology. Mine being *Ram*, I turn this black wheel now while holding the red one, to which a compass is fixed, so that Ram points to the direction, place or location I want to assess the Feng Shui of.

Before I demonstrate how the kit works, you need to know which elements are contained in every single reading. You have to follow through since if you leave out one, you will get a distorted answer. These elements are represented by *five rings* on the Lo Pan:

– 1st Ring: I Ching Hexagram

– 2d Ring: Direction

– 3d Ring: Element

– 4th Ring: Yin/Yang

– 5th Ring: Animal

So let's do a reading right here. I am sitting now here at my desk, in a newly furnished room that since I have entered this rented house, I have not used yet. My face points to the entry door and it is this direction that I would like to assess. So I adjust my compass so that the red needle is between the two red dots and thus points South. I look up the direction I want to assess and get the following information:

– First ring (I Ching sign): Chen

Now the book tells me that I need to know the other trigram so that I can look up the complete hexagram. In order to calculate the other trigram, the Kit explains, I need to make the following calculation (for males) which is called Your personal Pa Tzu Compass:

> **Man-Ho Kwok**
> Subtract the last two numbers of my year of birth from 100 and divide by 9.

My year of birth being 1955, I arrive at the number 5. Now I search for my personal compass and do not find it. I look further and see that people with compass number 5 should use compass number 2 if they are males. The trigram associated with compass number 2 being K'un I look up the combination of Chen (thunder) and K'un (earth) and arrive at the I Ching hexagram *24. Return*. The explanation reads at follows:

> **Man-Ho Kwok**
> This hexagram indicates a period of growth after a period of disorder and disintegration. It is a positive hexagram and is full of

possibilities. It is time for a fresh start so enlist the help of others, but do not try to rush change as it will take a natural course.

– Second Ring (Direction): East. Very unfavorable for compass No. 2. Since the entry of my house points to that direction, this could signify that I should not go out much, which actually I don't do. Culturally and socially, there is as good as nothing attractive for me here on *Lombok* island, except some nice beaches that, however, I do not frequent often since I am working very intensely on around forty publishing projects. On the other hand, the house I got here, a huge villa owned by an Arab merchant, is unique and represents in many ways what I have always dreamt of. So I really enjoy to be here and to have such a wonderful place for my work, too.

– Third Ring (Element): Earth. The book's explanation suggests to introduce plants or wooden products in that area. When I watch out of my entry door that my desk faces, I see the plants, bushes and trees of my front yard. So this is well adjusted. However, my personal element according to Compass No. 2 being Wood and according to the destructive cycle Wood destroys Earth, this reading is also partly negative.

– Fourth Ring (Yin/Yang): Yang. The fourth ring indicating Yang, I have to look up the seventh ring, too, which is related to my personal animal (Ram) and that indicates Yin. Yin and Yang are mutually complementary and balance each other out. So this reading is positive.

– Fifth Ring (Animal): Dragon. This ring has to bee seen in relation to the last ring which indicates the personal animal. As to the combination Dragon-Ram, the book explains:

Man-Ho Kwok

In this combination there is a temptation to let your imagination run free with design and color, but it may not always please others. You should try to be more open to advise.

This is true! I have behind me a phase of designing web pages and I wanted to impress my visitors with strong and bold colors. There was much red and orange. Eventually, after going on months and months that way, I got feedback from a good and close friend who told me he felt irritated and confused when looking at my pages. There was too much and too many colors so that at the end, he confessed, he turned away since he did not catch the message. What a change to have got his advise!

When I look back now at my former period, I cannot understand myself any more. How could I have liked that? But that's the way we change, *when* we change. That I should be more open for advise is equally true. Had I been more open to follow the advise of experts I would have avoided to add many unnecessary features in my web design. It would have saved me time and money to have followed this advise.

I admit that the readings you get using this Feng Shui Kit are somewhat superficial; but, of course, there are more erudite publications. Only, you have to have the time to really digest them and apply them in your life. To just fill your bookshelf with would be another waste of resources. So I take the essence out of the Kit's reading and complete it using my own intuition. That's a good combination, anyway!

Nancilee Wydra

The Feng: The Book of Cures

Lincolnwood: Contemporary Books, 1996

The Book of Cures is one of the practical Feng Shui books. The book, while it is compact in size, is more like a handbook or manual. And as it is often the case with such kind of books where the overwhelming part consists of looking up things (instead of reading the whole), the author is conscious about the fact that the reader needs some introduction in the general principles underlying the advice presented in the manual, before they can successfully apply that advice. And this introduction is excellent! It is among the best for those who desire a very short, very compact and very easy-to-read introduction into the highly complex science of Feng Shui.

I would go as far as saying that already the introduction (Part I) is worth buying this book. I myself do not need the practical advise or cures de-scribed in Part II, since I practice intuitive Feng Shui since many years, and even before I knew about this science.

Some glimpses, however, that I made into the Cures part of the book showed me that the advise is appropriate and written in a comprehensive manner that is easy to understand. I therefore concentrate upon this little jewel that the author may herself not consider as the major contribution this booklet makes: the *introduction*. It is divided into six chapters:

– Power of Place
– What is Feng Shui?
– Schools of Feng Shui
– The Five Elements

– The Pyramid School's Ba-Gua

– The Senses

This introductory part of the book is the most necessary to read for any un-initiated Western reader. Let me cite the way the author opens her book:

Nancilee Wydra

I can recall how places felt to me as a child. On my way to school each morning my girlfriends and I walked past our neighborhood's haunted house. The early morning sun peeping up from behind this home's conical roof created an ominous shadow on the cracked cement sidewalk.

Almost everybody can report similar feelings and observations from our childhood. But how many of us have noted them or inquire into their validity once being grown up? I have myself gone all the way from a materialistic worldview which was shaped during all those years of school terror until my today's open, holistic, dynamic and complexity-affirming worldview. It seems that only poets and very strong individuals are able to have their childhood intuition prevail over that cruel brainwashing school has done to most of us. I have fortunately gained it back by recovering and healing my inner child during a two-years hypnotherapy some years ago. Today I cannot understand how one can live with the dangerously reductionist worldview that is still the reigning consciousness paradigm of the majority of people in our times.

As Nancilee Wydra suggests, we can in fact dive into our childhood memories again and see what our knowledge was regarding power and places or energies that radiate from places or that is kept attached to objects. Her approach is interesting also from another point of view. She questioned in her book the immediate applicability of ancient concepts or concepts pertaining to certain cultures, in other cultures. For example, she cites ways of ill-interpreting Feng Shui rules because those rules fit a certain culture (the Chinese culture) with a certain set of behavior. She therefore suggests to abstract or extrapolate the underlying principle of every rule and then apply the extracted principle to the problem, and not the literal rule. This seems to be a wise approach, indeed!

I know from experience that it is exactly the lack of this capacity to abstract the general principle from ancient rules that make people today turn away from them. Dr. Joseph Murphy found the same to be true regarding the Bible truths. Applied literally, many of those truths would be absurd in our times or they would hurt people more than healing them. It is only through interpreting the ancient texts in the light of our today's psychological understanding that we can see the true and deep meaning of what is written.

To develop this understanding means to work with our intuitive and synthetic mind, using more of our right brain hemisphere's associative thinking capacity, than applying the strict logic ability of the left brain. This book is a step in the right direction also in this sense. After all, I can recommend this book especially to those who are new to the subject of Feng Shui, who are curious to learn the basics in a comprehensive way that takes a minimum of time.

Richard Craze

The Feng: Feng Shui Book and Card Pack

London: Thorsons, 1997

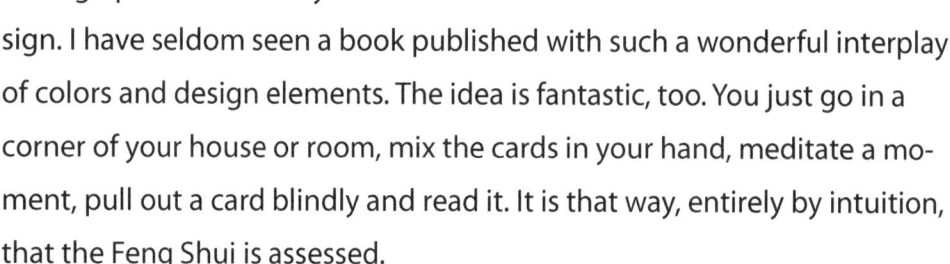

Feng Shui Cards are a beautiful item that convinces through perfect harmony between content and design. I have seldom seen a book published with such a wonderful interplay of colors and design elements. The idea is fantastic, too. You just go in a corner of your house or room, mix the cards in your hand, meditate a moment, pull out a card blindly and read it. It is that way, entirely by intuition, that the Feng Shui is assessed.

This intuitive approach to Feng Shui is so much the more satisfying for the beginner who has little knowledge yet of the intricacies and complexities of *Compass Feng Shui*. On the hand, of course, one must have a feeling for that magic reality - and some people are just not to turn on that way. So this booklet is probably not a mass selling item. So much the more I compli-

ment the courage and aesthetic sense of authors and publisher!

After a short introduction into the principles of Feng Shui, the booklet explains the approach to Feng Shui that is the most practiced now in the Western world: *Pah Kwa Feng Shui*, which uses a fixed arrangement of trigrams, derived from the I Ching, that are positioned in form of an octahedron and that overlays the map of a room or house on it. Logically, eight different sections will show up which are called *The Eight Enrichments*. Accordingly, for each of these eight main sections, the booklet shows Eight Remedies in case the card indicates some form of negative energy (the ch'i may for example be unpredictable or overpowering in one section).

Finally, an assessment of one's whole property according to the *Five Elements* is indicated before the booklet explains every card in detail.

The 32 cards themselves are divided in four different suites, according to the Four Animals that traditional Feng Shui teaching situates around every house or property: the Red Phoenix (South), the White Tiger (West), the Black Tortoise (North) and the Green Dragon (East).

To end this review of a short but wonderful booklet, let me cite the first sentence of the Introduction because this single sentence reveals the unique and powerfully holistic focus of the booklet:

> **Richard Craze**
>
> Taoism, which is the ancient religion of China, holds that what is, is. Unlike Christians, who believe in the heavenly paradise of an afterlife, or Hindus with their vast array of gods, or even the Buddhists' belief that all life is suffering and the only reward is in Nirvana, Taoists regard now as important and believe that there are no gods, heavens, or future paradises. Their heaven is order, harmony, balance and jen - love of life.

Apart from what this set represents in knowledge and practical value, it is also a wonderful gift item!

RICHARD GERBER, M.D.

A Practical Guide to Vibrational Medicine

Energy Healing and Spiritual Transformation
New York: HarperCollins Quill, 2001

A Practical Guide to Vibrational Medicine is an excellent book, very carefully written, very well put together conceptually, while I have to put a question mark regarding one conceptual matter that I will came back to further down, toward the end of this review.

Let me first comment on the general conception of the book and the author's unique contribution to a novel subject, and a very important one. Gerber's main quality is his detached and careful approach to a matter that, you

won't get the impression from the book, but really is very controversial. Let's not forget that a man like Wilhelm Reich, while perhaps a little unwise in some respects, had to die for introducing the energy approach to healing, and Paracelsus who was perhaps the first in our culture who came up with energy healing, had to stand trial against the Inquisition. Smart as he was, perhaps smarter than Reich, and perhaps surprisingly so, he did win the trial. But he had to appear in an humble costume, while in real life Paracelsus was not humble at all. Well, I would say that Dr. Gerber is quite the opposite character, and he rather understates, which adds to his credibility.

The overall impression I had from this book was excellent and my respect for the author is almost unlimited. It is not easy to write a book about such a controversial matter and yet write it from such a balanced, and yes, objective, perspective. I never had in this book the impression that the author wants to prove his point, or that he has something to defend. This is a character trait that surely many people dislike in similar productions, but here you can be reassured, you won't have a problem to read this book and your emotions will not get stirred up by unwanted polemics.

This being said, the book is perhaps not as practical as the title suggests it to be, not as practical as for example Donna Eden's book *Energy Medicine (1998)*. This is because this book is conceptual in the first place, and practical in the second place, and because it's paradigmatic, and cutting-edge in its overall perspective. It's well practical when you see the abundance of references and the intelligence with which the whole is put together. The resource section of the book also is very valuable, and indeed practical with pages and pages of organizations that can lead you further in your research project.

But the book from its overall style is a sound academic study, and when I say academic this is for me surely not a negative thing to note, but I know that for some people it is. The merit of this book is the vast research the author has done, and it can be considered as being a condensation of this research, in that it produces something like a synthesis of a lot of material that is only mentioned in the notes. For example, when you have heard of topics like acupuncture or Bach plants, or radionics for the first time, this book gives you a wonderful overview and introduction into each of these subjects. And then, in the notes and the reference part of the book you get the information you need for further research. A good idea of the author is also his *Recommended Reading* section (pp. 424-432), where he gives concise recommendations for further study. Very few authors have done that.

All this is extremely valuable as a package for you when you are a researcher and want to have an overview of the whole of the topic. Without a book like the present one, you start from scratch, as I did years ago, when there was not yet anything of the kind of this book, and when I really started with researching into the writings of Paracelsus, Mesmer, and Reich, reading them in their old German original, and then trying to work this information out in English for my self-help guides. And that's kind of frustrating because you are virtually overwhelmed by the abundance of the source material, and invest lots of time and energy without knowing if you ever will be able to draw real water from the sources.

What is also very strong in this book is how the author connects our modern perspective of vibrational medicine with the old teachings, the medical tradition of Antiquity, the esoteric knowledge of the Mystery Schools, Chinese medicine and acupuncture sources, or Chinese QiGong. Gerber writes:

> **Dr. Richard Gerber**
>
> Ancient approaches to understanding disease and body healing often viewed illness from the perspective of the human spirit, or the body's life-force energy. These somewhat mystical viewpoints may now hold the key to understanding why people become ill and how they can regain their health. Yet modern medicine tries to distance itself from ideas of spirit and life energy. Mainstream healers long ago gave up the belief system referred to as 'vitalism' or the theory of vital energy. But is vitalism really such an outdated concept when we begin to factor into the human equation some of the new discoveries in the field of quantum and Einsteinian physics that describe the underlying energetic nature of the physical world?/2

Here the author asks a very important and at the same time daring question. And I really welcomed it, as I always wondered why generations of scientists were able to brush off with this scant argument the most important

Published by Sirius-C Media Galaxy LLC, 2011

alternative scientific input, i.e. simply by stating that the alternative theory or concept was *vitalistic?* It needs courage to fight provincial and judgmental thinking, and Dr. Gerber has this courage, while he doesn't boast with it. I know that only an emotionally very mature person can write a book on such controversial matters as are exposed in this book, and Dr. Gerber is this person. This is another element in the value of the book. You could take it to court, so to speak, as a lawyer, for witnessing about the validity of energy medicine, so well it is written, and with such an authoritative and emotionally detached tone. This is exactly what we need to get ahead in our fight for a better world. I know how to defend such a case in court, but without a book like the present quoted in my expertise, I would get caught up in an unending series of research that I could never finalize, simply because I wouldn't have the time. And all starts with a sound definition of what in fact is this thing called *vibrational medicine*:

Dr. Richard Gerber

Vibrational medicine is based upon modern scientific insights into the energetic nature of the atoms and molecules making up our bodies, combined with ancient mystical observations of the body's unique life-energy systems that are critical / but less well understood aspects of human functioning. Rather than seeing the body as a sophisticated machine, animated only by electrochemical reactions, vibrational medicine views the body as a complex, integrated life-energy system that provides a vehicle for human consciousness as well as a temporary hosting for the creative expression of the soul./3-4

While traditional Western medicine never bothered about other than mechanistic and strictly causal, and linear, relationships in the etiology of disease, which is why it can be called a *reductionist* approach to healing, this is totally different with vibrational medicine:

Dr. Richard Gerber

Modern physics tells us that the only difference between these forms of energy is that each oscillates at a different frequency or rate of vibration. Hence, vibrational medicine refers to an evolving viewpoint of health and illness that takes into account all the many forms and frequencies of vibrating energy that contribute to the 'multidimensional' human energy system./5

Another important point of validation in the transition to a holistic model of medicine was human emotions, and how they were thought to impact on human health, or on illness. Gerber notes in a synopsis entitled 'Major Differences Between Conventional Medicine's and Vibrational Medicine's Worldviews' (p. 3) for emotions:

Conventional Medicine Model

Emotions thought to influence illness through neurohormonal connections between brain and body.

Vibrational Medicine Model

Emotions and spirit can influence illness via energetic and neurohormonal connections among body, mind, and spirit.

I would like to mention also how brilliantly, in a few sentences, Dr. Gerber describes the outdated mechanistic model, and why it's superseded in the first place. I have never found this elsewhere in this condensed form, except perhaps Capra's book *The Turning Point (1982/1987)*:

Dr. Richard Gerber

The concept of the body as a complex energetic system is part of a new scientific worldview gradually gaining acceptance in the eyes of modern medicine. The older, yet prevailing, view of the human body is still based upon an antiquated model of human functioning that sees the body as a sophisticated machine. In this old worldview, the heart is merely a mechanical pump, the

Published by Sirius-C Media Galaxy LLC, 2011

kidney a filter of blood, and the muscles and skeleton a mechanical framework of pulleys and levers. The old worldview is based upon Newtonian physics, or so-called billiard-ball mechanics. In the days of Sir Isaac Newton, scientists thought they had figured out all the really important laws of the universe. They had discovered laws describing the motion of bodies in space and their momentum, as well as their actions at rest and in motion. The Newtonian scientists viewed the universe itself as a gigantic machine, somewhat like a great clock. It followed, then, that the human body was probably a machine as well. Many scientists in Newton's day actually thought that all the great discoveries of science had already been made and that little work was left to be done in the field of scientific exploration./7

As I have mentioned often in my own publications, traditional medicine was vivisectionist in that it had to kill an organism before it would inquire in its functionality, thereby from the start dealing with a distorted view upon nature. Traditional medicine was studying death, instead of life, for gaining information about life, which could never result in a functional medical system. This distortion of science was mainly introduced by Aristotle's arrogant and in last resort absolutely stupid reductionism that, because it was later put on a pedestal by the Church, virtually annihilated anything even remotely correct in terms of nature observation. The result was that as Paracelsus reported in his books many more people were dying from the absolutely moron tyranny and incompetence of official medical practice rather than as a result of illness or old age, virtually in the blossom in their youth. While the Chinese, already thousands of years ago, had observed the living body, and never resorted to vivisection as a habit like Western doctors: the result simply is that Chinese medicine focuses upon health, and preventing disease, while Western medicine focuses upon illness, and how to prepare for death. That's why I pretend that Westerners never ever, at any point in history, had an even remote idea was *health* was; and this is equally true in their regard on emotions, and sexuality. Gerber writes:

> **Dr. Richard Gerber**
>
> While early European physicians could analyze the human body only in terms of dissection of organs at the time of autopsy, today's medical researchers have the tools to study our physical makeup at the cellular and molecular levels./7
>
> The old-world, Newtonian model of medicine lacks an appreciation of the seemingly intangible things such as emotion, consciousness, and the energy and life force of soul and spirit./9

So what are actually emotions in the regard of vibrational medicine? Dr. Gerber explains:

> **Dr. Richard Gerber**
>
> In the vibrational medicine view of human functioning, our emotions are not just the result of neurochemical reactions in the limbic system or the emotional centers of our brain. Our emotions are also influenced by a greater, spiritual energy field that encompasses and influences the entire physical body and nervous system./10

Wilhelm Reich once said that emotions are bioenergy in flow, energy in motion, which is why, as he explained, they are called emotions: as they are e-moted or moved out, squeezed out from the bioplasma. I think one can hardly express it in better terms. But this view was clearly marginal in the face of dull Western medicine that could in no way explain emotions other than by declaring them either of no importance, or of perverse importance. Now, as we are in medias res, and got the magic word mentioned, spiritual energy, let me focus on that important matter for a moment in asking *What is this spiritual energy?* Gerber doesn't brush over this subject, but in the contrary very carefully discusses the various points of reference, including the Asian tradition. But he ends up with different energies, a whole array of energies flowing through the human body, and not just one vital energy, one cosmic life force, and that really estranges me. Of course, this one energy, as early discovered as by Paracelsus, appears in many different vibrational octaves, such as sound appears in different octaves.

But for Gerber, there are different energies, to be precise, four, a chemical energy, an electrical energy, a ch'i energy, and a prana energy. Instead of seeing that different traditions named this one and only energy differently, the Chinese *ch'i*, and the Indians *prana*, he sets forth that ch'i was the 'acupuncture energy' and prana the 'chakra energy'. And as to 'chemical energy' and 'electrical energy', suffices to read Reich's explanations about the orgone to understand that chemical and electrical, as well as electromagnetic manifestations are an outflow of orgone, not different sorts of energy phenomena. I actually have asked the question, for a publication on Wilhelm Reich, to Mary Boyd Higgins, Director of the Wilhelm Reich Children's Trust, and Director of the Wilhelm Reich Museum in Maine, and she replied that I had well understood the fact that according to Reich's research, light, electricity and electromagnetic streams, on one hand, and orgone energy, on the other, were not different forms of energy but that orgone was the primal energy and light, electricity and electromagnetic streams were flowing out from that energy.

Well, I am not here to clarify any confusion that Gerber's new distinction might create, and I am not here to judge. Let us just note the fact that this otherwise very careful researcher and medical doctor came up with a view that obviously contradicts 5000 years of documented history of bioenergy. The future will show what is right here.

AMIT GOSWAMI

The Self-Aware Universe

How Consciousness Creates the Material World
New York: Tarcher/Putnam, 1995

What **Amit Goswami** can express poetically, in his book *The Self-Aware Universe*, not many can express it ever in words. But his poetry, to paraphrase Emerson, has an edge to it, otherwise it is none. The edge is quantum physics. Goswami's genius is that he's able to express very complex insights and relationships in a simple poetic language that even the commoner can understand. When I saw Goswami in *The Bleep*, I was already impressed by his unconventional yet powerfully convincing appearance, but when I read him line by line, it was an intellectual pleasure for me I seldom had when reading a science book.

While Goswami leaves no doubt that he defends the monistic paradigm in spirituality, which clearly means taking sides when you do this as a scientist, I respect it because he has justified his spiritual paradigm scientifically. I can say that Goswami's view of the universe sounds very coherent to me, and from his general style and reputation, this man is not a lighthearted spirit - pretty much to the contrary. And in *The Bleep* he was clearly underrepresented with only seconds of interviewing time granted, while others got plenty of minutes. And a minute, just one minute, in such an important movie is important.

This being said, this book surely is not an easy read. I had to fight through because mathematics never was my strong point, which is why I was thankful for Goswami's really wistful mix of mathematical and poetic explanations of his vision - and that is something unusual in our Cartesian science tradition - while it was not unusual in the Renaissance. And with this I wish to express a compliment: this book truly is well-written and the author has literary talent, which cannot be assumed to be the case normally with physicists, while we have been spoiled with Fritjof Capra because he set the standard of comprehensiveness very high! And Goswami got a sense of humor, too:

> **Amit Goswami**
>
> But we physicists are a stubborn lot, and we fear the proverbial toss of the baby out with the bathwater. We still lather and shave our faces watching carefully as we use Occam's razor to make sure that we cut away all 'hairy assumptions'. What are these clouds that obscure the end of the twentieth century's abstract art form? They boil down to one sentence: The universe does not seem to exist without a perceiver of that universe'./xiv

My own cousin is a physicist and he was telling me a tale about those *hairy assumptions* long ago; and he had about the same talent to explain complex matters in an easy-to-understand language. He was explaining me Einstein's relativity theory when I was just sixteen-years old. And I understood it, and from that time got a feel for the usefulness of physics, while school could not convey this to me. I think many of us are in a similar position in that we would like to enhance our scientific understanding without however having to digest volumes of mathematical gibberish, and pages of formulas. Capra and Goswami, and a growing number of other scientists today prove by their books that it does not need to be that way, and that knowledge, whatever level of complexity it assumes, is transmissible in ordinary language. At least *cum grano salis.*

And for describing the unique paradoxes of quantum physics, I haven't found an author who can explain them with similar ease:

Amit Goswami

Furthermore when it [the electron] is not a single particle it appears to be an undulating wavelike cloud that is capable of moving at speeds in excess of light speed, totally contradicting the Einstein concern that nothing material can move faster than light. But Einstein's worry is assuaged, for when it moves this way, it is not actually a piece of matter./xv

Goswami summarizes the *quantum paradoxes* as follows:

Amit Goswami

• A quantum object (for example, an electron) can be at more than one place at the same time (the wave property).

• A quantum object cannot be said to manifest in ordinary spacetime reality until we observe it as a particle (collapse of the wave).

• A quantum object ceases to exist here and simultaneously appears in existence over there; we cannot say it went through the intervening space (the quantum jump).

• A manifestation of one quantum object, caused by our observation, simultaneously influences its correlated twin object - no matter how far apart they are (quantum action-at-a-distance)./9

Goswami is not one of those physicists for whom Einstein is the living one-and-holy God. He shows that Einstein's speed of the light limitation is none when applied to subatomic physics since we are dealing not with matter, but with waves, thus contradicting other physicists who speak in this case about *exceptions from relativity theory*. No, the wave behavior of electrons doesn't represent an exception from relativity theory as relativity clearly applies for matter only, for mass, and not for energy, for waves. Goswami explains:

Amit Goswami

According to quantum physics, even though the two electrons may be vast distances apart, the results of observations carried out upon them indicate that there must be some connection between them that allows communication to move faster than light./xv

In a similar mood and with the same eloquence, Goswami explains why we need to overcome the Cartesian split:

Amit Goswami

Since René Descartes divided reality into two separate realms - mind and matter - many people have tried to rationalize the causal potency of conscious minds within Cartesian dualism. Science, nevertheless, presents compelling reasons to doubt that a dualistic philosophy is tenable: In order for the worlds of mind and matter to interact, they must exchange energy, yet we know that the energy of the material world remains constant. Surely, then, there is only one reality. Here is the catch 22: If the one reality is material reality, consciousness cannot exist except as an anomalous epiphenomenon./10

With the same lucidity, Goswami discusses at length, and eventually rejects material realism as a foundation for any kind of holistic science of the future:

Amit Goswami

The negative influence of material realism on the quality of modern human life has been staggering. Material realism poses a universe without any spiritual meaning: mechanical, empty, and lonely. For us - the inhabitants of the cosmos - this is perhaps the more unsettling because, to a frightening degree, conventional wisdom holds that material realism has prevailed over theologies that propose a spiritual component of reality in addition to the material one./11

What many people ignore, in fact, is that quantum physics did not per se establish a holistic science paradigm. Capra has discussed this question in *The Turning Point*, pointing out that quantum physics is restricted to the subatomic realm, while in conventional physics the Newtonian mechanics is still valid. Goswami explains:

Amit Goswami

The philosophy of materialism, which dates back to the Greek philosopher Democritus (ca. 460 - ca. 370 B.C.) matches the worldview of classical physics which is variously termed material, physical, or scientific realism. Although a new scientific discipline called quantum physics has formally replaced classical physics in this century, the old philosophy of classical physics - that of material realism - is still widely accepted./15

This is why, as Goswami discusses at length, the mere decision pro or con quantum physics does not change much in the landscape of physics. What does this change, Goswami says, is the philosophy behind the screens. And here he points out with many examples how physics is shaped by the underlying spiritual or non-spiritual paradigm. He forwards a catchy comparison from the brain-mind discussion to draw a parallel:

Amit Goswami

Classical functionalism assumes that the brain is hardware and the mind software. It would be just as unfounded to say that the brain is classical and the mind quantum. Instead, in the idealist model proposed here, the experienced mental states arise from the interaction of both classical and quantum systems./173

It seems that Goswami's choice of the idealist model and philosophical monism was not just the result of cultural conditioning. As he explains, and as it is well-known, India in the whole of its philosophical tradition adhered to spiritual monism, and idealism. But the strength of Goswami's coherent

Published by Sirius-C Media Galaxy LLC, 2011

view of modern physics is that he carefully double-checked the results of all the various philosophical constructs, in their effect on scientific observation at the quantum level. On the other hand, his clear choice of a spiritual direction may interfere in some ways with his scientific objectivity. It is quite difficult to see this as a non-physicist but as a researcher I find his bias a little bit too strong. While I profoundly respect and admire the Indian spiritual tradition, when a quantum physicist makes such a spiritual choice as a base paradigm also for his research, I must question his objectivity.

I will stop my comments here in the hope that this information is sufficient to raise your interest in the present book, and take the challenge to read it. While it is written in a good conversational English and without too much science gibberish, a robust and lucid ability to follow convoluted and complex philosophical, and to a lesser extent, mathematical explanations is required.

VALERIE HUNT

Infinite Mind

Science of Human Vibrations of Consciousness
Malibu, CA: Malibu Publishing, 2000

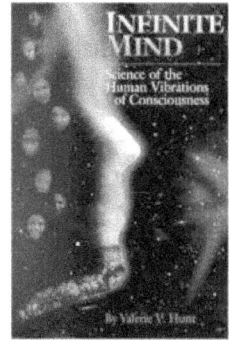

I found *Infinite Mind* by

Valerie Hunt only recently, while it's by far not a new production. It is published in 2000, but the research it is based upon dates back to the 1970s. But that does in no way turn down or diminish the importance of this book and its underlying research. In the contrary, it shows that every thorough and profound research needs decades to really condense into something that we call a science. I know this from my own experience as a science author; the research for my books on the human energy field started out more than twenty years ago. And then, there is another lapse of time involved in this science to be recognized by the established science tradition and academia! This has to my knowledge not been done yet specifically for Hunt's *Science of the Human Vibrations of Consciousness*, but it has been done in a larger framework, within what we call today *Consciousness Research*, and which has been fertilized by many different sciences. And within this larger framework, Hunt's approach to consciousness boosting is no more a revolutionary personal attempt of moving beyond the limited borders of the *Cartesian Worldview* that it seemed to be when Hunt started out in the 1970s, much against the stream! It is today something like established knowledge, and yet known only to a tiny elite of avant-garde scientists and intellectuals worldwide.

Considering this early research I would like to focus in this review on those phenomena that the author observed, and measured, or that she reports having been measured by others, and that later, in some cases only very recently, have been corroborated by further research.

There is a staggeringly simple experiment that was repeated over and over and where observations coincided over time, and with various researchers. It is an advanced form of guessing or intuiting answers that usually is done with a computer and where the test person clicks the mouse or hits a pad every time, and as fast as possible, to give the answer to a specific question. The author writes:

Valerie Hunt

We observed that before the brain wave was activated and before stimuli altered the heart rate, blood pressure or breathing, the field had already responded. This led us to postulate that a person's primary response in his world takes place first in the auric field, not in the sensory nerves nor in the brain./25

The same results for the same kind of experiment was reported by Dean Radin and Michael Talbot, in their respective books, reviewed in this volume. Today the measurements can of course be much more precise compared to the 1970s, as Dean Radin reports. Errors can be as good as excluded. It was clearly shown in these experiments that the field reacted long before the stimulus was getting to the brain, let alone triggered the response of the hand and arm muscles to do the actual clicking of the mouse or hitting of the pad. This important experiment shows that in many responses in life, especially in vital ones, but also to a certain extent in intellectual responses, a pre-intellectual decision making takes places that is totally intuitive, and that can be located not in the brain, because it's precognitive, but in the aura, energy field or etheric body of the person. What answers here is thus not the brain, not the cognitive apparatus, but the field itself, the bioenergy. To use my own terminology, the answer is given by E, the supreme intelligence, through its energy, the e-force which is what composes our aura and our bioplasma.

The next important point to consider is the relationship between the observer and the observed, that equally was extensively treated in the studies of both Dean Radin and Michael Talbot, and countless other more recent studies. Valerie Hunt writes:

> **Valerie Hunt**
>
> A subatomic interpretation is that there is no universe without an observer. It has been said that there is no physical universe without our thoughts about it. Quantum physics reminds us that the moment one inquires into matter, like an electron that has no position, velocity, momentum or spin, that electron acquires character. Simply, we cannot observe the world without participating with it. Observers are part of the nature of physical reality, where matter and mind blend. Furthermore, when studying open dynamic systems, there can never be identical answers. The importance of repeating studies is not to determine 'truths', but to disclose many truths - different pieces of information to fill in the puzzle./44

But perhaps the most important research is the one done directly on the field, research that formerly was called *aura research* and that we now call research on the human energy field. There are references in both Dean Radin and Michael Talbot that lead to further references, and this research is so vast today, branches out so abundantly that my guess is it will be the foremost research topic in the future. It will probably open the door to our passing way beyond the speed of the light and allow us to build magnetic-driven spaceships, as we know them from science fiction. Besides, the applications in daily life are so countless that I do not even mention them here. The author writes:

Published by Sirius-C Media Galaxy LLC, 2011

Valerie Hunt

Not until we investigated practices of Eastern medicine and acupuncture did we give serious attention to human energy fields. Still, Western science does not consider the human auric field a credible area for research. If one cannot see the aura and discussion of it is couched in unfamiliar language from other cultures, one doubts its value. Ancient writings claiming that chakras are the auric field source with meridian pathways the circulation route do not fit snugly into the current understanding from structural anatomy. Nonetheless, the few who have chosen to research this uncharted human field area discover facts unique to living fields that also correspond to universal laws. The human field looms as primary to life./65

As this book is very large and complex, I will focus here further on Chapter VI of the book entitled 'Emotions: The Mind-Field Organizer' because I believe this is one of the most important topics of the book and because I have myself done extensive research on emotions. So I will just list the quotes here that concern Hunt's alternative view on emotions, which exactly coincides with what I have written in my own books on the subject, without knowing about Hunt's research at the time when I wrote my own books. At the next update of my books I will insert the quotes, and discuss Hunt's approach in detail. It is the first time in my twenty years of research on emotions that I meet a mind who is broad and smart enough to really understand the energy nature of emotions:

Valerie Hunt

Emotions carry the essence of our unique and collective consciousness. (...) I suggest that human emotion is the organizer of energy fields./104

Current ideas about the psychology of emotion need to be re-evaluated./105

In the last 25 years, 100 new schools of psychology have been established. But there has not been a revolutionary new idea about human emotion since the early part of this century. While it is true that transpersonal psychology is pointing to higher spiritual aspects of consciousness, it is still with a weak voice that is not commanding the attention paid to the old models. Actually, even transpersonal psychology is not radically new; it is merely an extension of older concepts./106

I have done this re-evaluation of emotions research in my books and I speak of a unique *emotional identity code*, and now find in the present book, that I found only after finalizing my books, that Dr. Hunt speaks of a specific personal emotional field signature (p. 111), which she describes as a form of *steady state* of emotionality that represents something like a unique emotional patterning that differs from one person to the other, if I understand it correctly. Dr. Hunt writes:

Valerie Hunt

In contrast to the emotions at the material body level, in altered states there is evidence of an open emotional system that is dynamically in touch with deep needs and subtle happenings in the universe. Here in the no-time/space realm, one discovers free emotional energy, a super-consciousness state, the home of the peak experiences that we never forget. Here the closed system opens, revealing a broad continuum of emotions that explain things we knew about ourselves, particularly the schisms in our awareness./106

I strongly believe that the internal dynamics of the most complex biofield, the human energy field, are based on its emotional organization./109

At the deepest level, all things are composed of vibrations organized into fields that permeate the entire structure. Fields, /

Published by Sirius-C Media Galaxy LLC, 2011

whether biological or otherwise, have their own integrity. They are organized, not random, and they have the capacity to selectively react, interact, and transact - to respond passively, and to cooperatively unite with other fields. In other words, the mind aspect of the field, the aspect with the highest vibrations dynamically guides all choices and transactions as it influences and is influenced by all other fields./109-110

In my research on emotions and also my even more extended research on *shamanism* and what I came to call the *eight patterns of living*, I have started my scientific journey exactly with a closer look at what is *pattern* in nature, and how pattern is distinct from principles, as we know them from our scientific formulations of natural laws.[5] Dr. Hunt writes:

Valerie Hunt

Patterns of the mind dictate complex human behaviors; brain patterns activate simpler ones. Every experience has concomitant emotions, and every emotion temporarily restructures the field. Activated emotions increase the electromagnetic flow of the field. Likewise, emotions arise from an altered electromagnetic environment./110

My research shows that human energy fields display a continuum. The extremely low frequencies (ELF), are directly involved with life's biological processes. The extremely high / frequency (EHF) patterns ally with the mind-field and awareness. The general pattern of ELF is similar for all people, while the EHF reveals a personal signature of emotional patterning for each person. Therefore, an individual's mind-field patterns may show unique clumps of energy at different frequencies with breaks in the frequencies along the total mind-field spectrum./110-111

---

[5] See, Pierre F. Walter, *The Idiot Guide to Emotions (2010)*.

I will end my review here while this book contains much more highly interesting and partly novel material, but it would go beyond the limits of a book review to cover all the subjects treated in the book. I recommend this book to every researcher on systems theory, on living systems, on the bioenergy, and also particularly to those who have written extensively about the orgone research by Wilhelm Reich and think, in some other kind of scientific reductionism Reich was the only one in the universe who has researched on the cosmic energy and the unique energy fields that are biogenic, evolutionary and informational and that really make out the *Secret of Life*, as Georges Lakhovsky entitled one of his books on the subject.

Published by Sirius-C Media Galaxy LLC, 2011

SHAFICA KARAGULLA

The Chakras

Correlations between Medical Science and Clairvoyant Observation
With Dora van Gelder Kunz
Wheaton: Quest Books, 1989.

This is an extraordinary book. I do research on the bioenergy since almost two decades, and I have not encountered so much information about such esoteric a subject in one single book. But I must warn the non-scientific reader: this is not a book for enhancing your general knowledge about the aura, and the chakras, and it is by no means a practical book, guidebook, or anything of that kind. You have got two medical practitioners here, one of which, *Dora van Gelder Kunz*, is a clairvoyant. The author herself, Shafica Karagulla, is the kind of traditional physician who writes with a lot of 'faculty terms', so to speak, using medical terminology all over the place. Further down, I'll quote some examples. So think twice if you want to buy this book. For me, it was indispensable for my research. There are some elucidations in this book that I found earlier in my research, but only after studying tedious manuals and old hermetic writings. One detail also is important somehow. Dora van Gelder died before this book was even in a draft, and therefore Karagulla was not always sure when she gave detailed accounts on Gelder's paranormal perceptions. This is obviously a bad fate, as part of the theory rests on assumptions. On the other hand, from her memory, Karagulla could relate many an anecdote about the powerful personality of van Gelder and her lucid intelligence. One thing she relates to have been a constant in van Gelder's sayings was:

Shafica Karagulla

There is nothing 'supernatural' in the universe; whatever phenomena appear so to us are the result of our ignorance of the laws that govern them./5

And we are reminded of the German poet and scientist Johann Wolfgang von Goethe, himself an initiate, who said that all secrets of the universe could be known to the common man, if only he could free himself from *school wisdom*, which was the eternal parody of real knowledge. This being said, there truly is precious knowledge contained in this book. In the following three sentences alone contain more than a whole library of non-scientific 'esoteric' samplers full of assumptions and half-truths:

Shafica Karagulla

From clairvoyants we learn that the personality includes three types of energy fields - the etheric or vital, the astral or emotional, and the mental - all of which surround and interpenetrate every cell of the physical body. The interplay among these three fields may be likened to what a musician calls the major chord, which is composed of three frequencies that in combination with four other notes form an octave of seven frequencies. It is said by some that every human being emits a unique tonal pattern which is created by his individual energy fields working in unison. This is sometimes referred to as the personality note./2

This elucidation is precious in the present scientific debate about what the cosmic life energy really is and especially the question if it is one single energy, or several energies. I have discussed in my review of Dr. Gerber's book *A Practical Guide to Vibrational Medicine (2001)* that I found it highly confusing that the author does not speak about one energy, but several, and even sees *ch'i* and *prana* as different energies. First, Karagulla speaks not of energies, but of energy fields. Second, the etheric, astral and mental fields have been recognized since times immemorial as different densities of the cosmic energy field. The etheric field is the densest, the mental field the most transparent energy field. But we are speaking here by no means about different energies! However, shortly after this important elucidation, Karagulla falls in the same trap and assigns to electric and magnetic fields the character of 'energies':

> **Shafica Karagulla**
>
> This and other research points significantly to the fact / that in addition to the electrical and magnetic fields which surround all physical events there are other types of energies and frequencies that are as yet undetectable by any instrumentation so far developed. For this reason, the gifted human being is the only 'instrument' so far available or the kind of research which correlates clairvoyant perceptions with medical diagnoses./5-6

While my point of view surely is not authoritative, I am convinced that in accordance with Reich's observations on the orgone, the electric and magnetic fields are *manifestations* of the primal cosmic energy, and not different energies. The following quote may point to a similar interpretation. If we can admit a universal field, as it has recently been done, for example through the research of Lynne McTaggart, exposed in her brilliant book *The Field (2002)*, then we are back at the ground, and can affirm there is only one field or energy, and not a confusing mess of perhaps conflicting energies:

> **Shafica Karagulla**
>
> This growing perception of the interrelatedness of all living things has many implications. For our purposes, however, we focus on the fact that there is a continuous energy exchange between the individual and the environment which every living system (whether human, animal, vegetable, or even chemical) regulates in terms of its own self-organization. This energy exchange is so constant and so indispensable for all living organisms that it can be regarded as a universal field effect./12

The notion of fields is introduced to every school child who is exposed to the experiment of proving that when electricity flows through a wire it establishes a magnetic field. Other physical field phenomena can also be easily demonstrated, although detection of the nuclear fields requires more sophisticated equipment. But when we posit a universal life or vital field (as we do in this book), this is much more difficult to demonstrate in a tangible way, for there are as yet no scientific instruments capable of detecting the presence of such a field. Yet life, if as yet undefined, is real, and the living (as open systems) have specific characteristics not shared by inorganic matter. The most important of these is the ability to replenish energy (what we call vitality) without an outside agency, which no machine can do./13

Another important detail in the research for the present book was the authors' focus on energy patterns. I have seen in my own research on the bioenergy that we can establish as fact the observation that life is coded in energy patterns, and not in any form of 'matter' as a primary substance of creation.

Shafica Karagulla

At a more fundamental level of physical being, we are becoming accustomed to thinking of ourselves in terms of systems, processes, and patterns of energy, rather than of dense materiality./20

In the view being developed here, man is a system of interdependent force fields, within which energy patterns are not only appropriate to the particular field but are also ordered by special processes and mechanisms. Furthermore, these energy patterns are responsive to changes in consciousness, a fact which gives us a very different perspective upon many of the troubling problems of human life./26

Published by Sirius-C Media Galaxy LLC, 2011

Now, going in medias res of this book, we have to ask the pertinent question how, once we know that all disease is a result of either lacking or misdirected bioenergetic flow in the etheric body, we can balance the bioenergetic setup so as to bring about health? Karagulla notes:

> **Shafica Karagulla**
>
> We found that abnormalities observed in the major etheric chakras were an indication of a tendency to a disease process, and that the area in which this would occur could be predicted even years before the symptoms began to manifest./6

The author's focus on what is health, first of all, or more specifically so, what is vitality, and reach the conclusion that it's a state of energy:

> **Shafica Karagulla**
>
> Vitality per se is not is not recognized as a form of energy in the West, but in the East, where it is known as prana, it has always been perceived as a universal force in nature connected with breathing and breath./28

How is this energy being supplied and replenished in the organism? Do we as yet have all the information we need? The author honestly admits that not all is known here, but we can be sure that in matters of replenishment, the direction is from the subtle to the dense, from the ethereal to the material, from the higher energy level to the lower, and not vice versa:

> **Shafica Karagulla**
>
> The etheric body vitalizes the physical body, but exactly how this takes place is not yet known. Since the / etheric channels run parallel to the nervous system, however, there may be a process of induction./28-29

As here only one energy is mentioned, prana, we have a further stone in the puzzle that speaks for these authors affirming that there is only one cosmic life energy, and not several. Now, how does this energy body, this aura, look like? Karagulla describes it:

Shafica Karagulla

To the clairvoyant, the etheric body looks like a luminous web of fine bright lines of force which, in a healthy person, stand out at right angles to the surface of the skin. Its texture may be fine-grained or coarse, a characteristic which repeats itself in the physical body type. Each organ of the body has its etheric counterpart, through which the etheric energy circulates constantly./30

The color of the etheric body is a pale blue-gray or violet-gray, slightly luminous and shimmering, like heat waves above the earth on hot days. In the average person it extends from five to seven centimeters (two to three inches) beyond the periphery of the physical body, gradually fading away into the enveloping ocean of etheric energy. This ocean of energy is in constant rapid motion, and surrounds the body much as the atmosphere surrounds the earth./30

And how does chronic illness look like? What is the imprint it makes in the luminous body and how can these imprints be identified. Here again, a clear answer is provided in the book:

Shafica Karagulla

It should not be supposed, however, that the existence of this etheric web inhibits the normal interaction between the emotional and etheric fields. In a healthy individual, there is an ordered relationship and a rhythmic flow among all the energy fields. But when there are chronic disturbances on the emotional level, such as continuous hostility or anxiety, the energy discharge is disordered, and this can eventually damage the whole system. To take another example, fear and depression tend to cut down the normal flow of energy, so that organs like the kidneys become less able to function normally. Thus the emotions closely affect both the etheric and physical bodies./31

Now, more specifically, what is the role of the chakras? Usually, from self-help books we learn only about the seven ordinary chakras, but there is more to it, as there are also chakras in both the emotional and the mental bodies, so in total there are not 7 but 24 chakras. The authors explain:

Shafica Karagulla

The seven etheric chakras, which are so influential in the health of the physical and etheric bodies, have their counterparts on the astral and mental levels. Like the physical body, which is continually disintegrating and rebuilding itself, the etheric, emotional and mental fields are constantly changing, but at a much more rapid rate. The chakras are involved in this change./34

What the chakras do is basically to transmit and transform energy, and their mechanism 'synchronizes the emotional, mental and etheric energies' (41). You may know from more popular books that paranormals see 'colors' in the aura. What does that mean? Can the colors be associated with certain characteristics? The authors provide stunning information here:

Shafica Karagulla

The colors, which vary from chakra to chakra, also glow in a way that contributes to their flower-like appearance. In a healthy person, the chakras' forms are beautifully balanced, symmetrical and organic, with all the parts flowing together in a rhythmic pattern. Their motion is, in fact, harmonic or musical in character, with rhythms which vary according to individual, constitutional and temperamental differences./35

The energy pours in through the core of the chakra, reaches the spine via its stalk, then flows along the tiny pathways of the etheric body which are connected with the physical nervous system. It finally returns to the chakras, moving outward in spirals through the periphery of the petals, in a constant intake and outflow./36

What is perhaps quite unexpected is that a simple clairvoyant regard on the chakras can reveal much about the spiritual evolution of the person, and their level of consciousness:

Shafica Karagulla

The chakras also reveal a person's quality of consciousness and degree of personal development and abilities, through the variations of the etheric centers and their interconnections with those at other levels. In a simple, rather undeveloped person, the chakras will be small in size, slow in movement, dull in color and coarse in texture. In a more intelligent, responsive and sensitive person they will be brighter, of finer texture and with a more rapid movement, and in an awakened individual who makes full use of his powers, they become coruscating whirlpools of color and light./36

Another interesting parallel is that the chakra network in the organism bears a resemblance with the endocrine system:

Shafica Karagulla

Certainly the intricate relationships among these chakras, as well as those on other levels, bear a close resemblance to the functional interconnectedness of the endocrine system. In fact, the interaction of all the fields with the physical body is a beautifully integrated system which originates in and is sustained by the energy patterns of the chakras in the etheric, astral and mental vehicles./37

Published by Sirius-C Media Galaxy LLC, 2011

This following concise little summary of the main functions of the chakras is hardly to be found in any other book:

Shafica Karagulla

To sum up, the principal functions of the etheric chakras are to absorb and distribute prana or vital energy to the etheric body and, through it, to the physical body, and to maintain dynamic connections with the corresponding chakras in the emotional and mental bodies. One of the functions of the chakras is to co-ordinate the interaction among the various fields. The condition of the physical body is affected not only by the rate of etheric energy flow, but also by the degree of harmony in its rhythm, and any obstructions which deform the normal energy patterns result in loss of vitality and ill health./38

I leave out from this review what the authors write about each chakra, as this information would render this book review too extensive. Let me state here only that the heart chakra is of particular importance as it is exactly in the middle between the upper chakras and the lower chakras. Thus, it bears a particular importance for balancing the energies. The authors point out:

Shafica Karagulla

In meditation, the student is encouraged to focus on the heart center, in order to strengthen its connection with the core of the crown chakra. This brings about a state of true balance in the body, for the heart center is really the point of integration in the whole chakra system, and therefore has an important overall balancing effect./42

More than twenty years ago, I found in one of Wilhelm Reich's books the surprising statement that *emotions are energy*. As I looked around, I saw that Reich, at his time, was quite the only medical doctor, scientist and psychiatrist who was saying this. I was intrigued and began a long research on emotions. But I could hardly find anything but the notorious assumption that emotions were 'difficult to grasp research objects' and that their nature was little known, while in the esoteric literature it was always assumed that emotions were related to the vital energy. Now, this book gives conclusive evidence to the energy-based nature of emotions, and generally, the *emotional field*. This is also highly important: the emotions are by no means 'in the brain', but flow pretty much like electric currents in the emotional body, which is the second subtle body we carry around our physical body. Karagulla explains:

Shafica Karagulla

Thus the astral or emotional field is truly universal. It is a fluid world of fast-moving images, shimmering with color and full of symbols and images that move us with their beauty or fill us with fear and anxiety, since it can be responsive to false and negative ideas as well as to those which are noble and uplifting. But in every case the emotional field is an intrinsic component of human life which needs to be understood and appreciated for what it is. (...) The emotional field is permeated by energy, as are the physical fields, but in this case it is moving much more rapidly, and is therefore perceived as a higher octave of color and sound. The form of the individual emotional field (the astral body or aura) has certain structural / features which correspond to those of the etheric field and the physical body itself. To the clairvoyant, this structure appears as a multicolored aura extending thirty-nine to forty-five centimeters (fifteen to eighteen inches) beyond the physical body. It looks rather like an ovoid, luminous cloud surrounding the body, as though the individual were suspended inside a semitransparent bubble of changing colors and patterns./48-49

Now, after we know what emotions are, and where they are located, let us look at what they do:

Published by Sirius-C Media Galaxy LLC, 2011

Shafica Karagulla

The activity in the field of the emotional body can be compared to changing conditions in the atmosphere of the earth, when observations from weather satellites verify / the areas in which storms are raging. In much the same way, the clairvoyant can perceive the emotional storms which trouble an individual as they disturb the aura./50-51

In our immediate neighborhood, we people our space with our emotional images, whether positive, negative or neutral./51

When we interact emotionally with others, there is an energetic flow involved in this communication:

Shafica Karagulla

If we accept the idea that we are dynamic systems which are constantly receiving and radiating energy, we can understand the degree to which human beings affect one another's emotional fields. This varies, of course, according to the inner stability and integration of the individual. When a person identifies himself with his emotions, he naturally responds readily to the emotions of others. He may be a warm and loving person, but he may also become the victim of other people's emotional disturbances./51

There is another very interesting parallel with Reich's discoveries about emotions. Reich was talking all through his books about the *emotional plague* as a major pathological development in humanity that he thought was caused by non-handling or mishandling our emotions through moralistic life-denial. Now, what Karagulla says is basically the same:

Shafica Karagulla

Over the years, humanity has produced a great deal of smog or debris in the emotional atmosphere./51

The importance of emotions as they are seen and evaluated by van Gelder and Karagulla by far surpasses even the most avant-garde research about emotions, to name only Candace B. Pert's bestselling book *Molecules of Emotion (2003)*. From what they say, in fact we can conclude that our idea of the *rational mind* is a pure fiction, because our mind is constantly connected with our emotions:

> **Shafica Karagulla**
>
> Because the mental and emotional fields are so closely interconnected, the mind is colored by emotion, just as the feelings are conditioned by thought. This is a universal characteristic, but when it is unbalanced or out of control the condition may become pathological. However, if the mind is not hampered by emotional stresses, it is a fine and flexible instrument for integrating and assimilating all levels of personal experience: mental, emotional and physical./59

From here, it's but step to bring forward a more general theory about the interaction between mind and brain, which is one of the most important topics of current neurological and psychoimmunological research:

> **Shafica Karagulla**
>
> The view of the mind/brain relationship which emerges from our research is very different from that generated by most psycho-physiological theorizing. Far from being a product of brain activity, the distillation of meaning and the interpretation of experience are seen to derive from a deeper level of the self. Such insight is then developed rationally by the mind and related to other knowledge, while the brain, which is the mind's instrument or physical partner, registers the information. In other words, the mind is dependent upon the brain for physical expression, but it also transcends the brain mechanism and can to some extent compensate for its defects./60

Published by Sirius-C Media Galaxy LLC, 2011

To come back to our initial question: how can we identify the presence of a pathological development through contemplating the energy patterns in the aura? According to the authors, it's a matter of how the energy is organized:

Shafica Karagulla

When the energy pattern is closely knot, it is very resistant to invasion from the outside world, but when loose and porous it can be penetrated more easily, and therefore the subject is apt to take in whatever may be in the / surrounding environment./92-93

Now, to come to an end of this rather large book review, I would like to emphasize that the few quotes I have replicated here in this review are by no means representative for the whole book. In fact, they are rather the exception from the rule. The most part of this book is written in medical terminology and thus not easily accessible to the novice reader. Yet despite this limitation, which is not really a limitation, this book is a real jewel in every new science library, and it probably will be discovered again and again, as it seems to me that so far, this book has not been given the scientific attention it deserves.

GEORGES LAKHOVSKY

The Secret of Life

Kessinger Publishing, 2003
Originally published as *Le Secret de la Vie, Les Ondes cosmiques et la Radiation vitale*
Paris: Gauthier-Villars, 1929

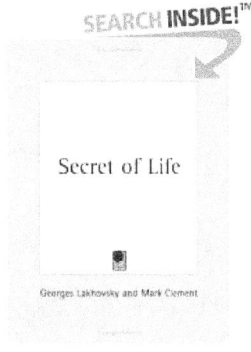

Back in 2004, I bought George Lakhovsky's books in their French original, the early original editions, and began to translate quotes from them. These books are precious stones in every science library, to be true, and the knowledge they bring is not theoretical, but very practical. Georges Lakhovsky (1869-1942) was not an academic, he was a practical man, a Russian engineer who emigrated to France before World War II. Lakhovsky found that all living cells possess attributes that normally are associated with electronic circuits. From this starting point and the observation that the oscillation of high frequency sine waves when sustained by a small, steady supply of outside energy of the right frequency would bring about what Lakhovsky called *resonance*. He arranged experiments showing that living cells respond to oscillations imposed upon them from outside sources. This outside source of radiation was attributed by Lakhovsky to cosmic rays that constantly bombard the earth. On the basis of these insights, Lakhovsky construed devices for healing by the application of high frequency waves, that today we know as *radionics*.

Lakhovsky saw that when outside sources of oscillations are resonating in synch with the energy code of the cell, the cell's growth become stronger, while when frequencies differ, this weakens the vitality of cell. From this primary observation, Lakhovsky further found that the cells of pathogenic organisms produce different frequencies than that of normal, healthy cells.

Lakhovsky specifically observed that if he could increase the amplitude, but not the frequency, of the oscillations of healthy cells, this increase would dampen the oscillations produced by disease causing cells, thus bringing about their decline. However, when he rose the amplitude of the disease causing cells, their oscillations would gain the upper hand and cause the person or plant to become weaker and more ill.

As a result of these observations, Lakhovsky viewed the progression of disease as essentially a battle between resonant oscillations of host cells versus oscillations emanating from pathogenic organisms.

He initially proved his theory using plants. In December, 1924, he inoculated a set of ten germanium plants with a plant cancer that produced tumors. After thirty days, tumors had developed in all of the plants, upon which Lakhovsky took one of the ten infected plants and simply construed a heavy copper wire in a one loop, open-ended coil about thirty centimeter (12″) in diameter around the center of the plant and held it in place. The copper coil was found to collect and concentrate energy from extremely high frequency cosmic rays.

The diameter of the copper loop determined which range of frequencies would be captured. Lakhovsky found that the thirty centimeter loop captured frequencies that fell within the resonant frequency range of the plant's cells. This captured energy thus reinforced the resonant oscillations naturally produced by the nucleus of the germanium's cells. This allowed the plant to overwhelm the oscillations of the cancer cells and destroy the cancer. The tumors fell off in less than three weeks and by two months, the plant was thriving. All of the other cancer-inoculated plants, those that were not receiving the copper coil, died within thirty days.

Lakhovsky then fashioned loops of copper wire that could be worn around the waist, neck, elbows, wrists, knees, or ankles of people and found that over time relief of painful symptoms was obtained. These simple coils, worn continuously around certain parts of the body, would invigorate the vibrational strength of cells and increased the immune response which in turn took care of the offending pathogens. Upon which Lakhovsky construed a device that produced a broad range of high frequency pulsed signals that radiate energy to the patient via two round resonators: one resonator acting as a transmitter and the other as a receiver. The machine generates a very wide spectrum of high frequencies coupled with static high voltage charges applied to the resonators. These high voltages cause a corona discharge around the perimeter of the outside resonator ring that Lakhovsky called *effluvia*. The patient sat on a wooden stool in between the two resonators and was exposed to these energies for about fifteen minutes. The frequency waves sped up the recovery process by stimulating the resonance of healthy cells in the patient and in doing so, increased the immune response to the diseased organisms.

George Lakhovsky's research demonstrates that it is experimentally possible to manipulate the resonance pattern of a cell's energetic vibration in order to actively fight a cancerous tumor, thus gradually eliminating the cancer. The special note of this treatment is that contrary to present-day cancer treatments, it changes the disease by changing the energy pattern of the affected cells, and not by chemically interfering with the cell's metabolism.

Lakhovsky called the cosmic life energy *universion*, which is the title of one of his lesser known books. I have called this bioplasmatic energy *e-force*.[6]

---

[6] See, *Walter's Encyclopedia, Academic Edition (2010)*, under 'Emonics'.

Published by Sirius-C Media Galaxy LLC, 2011

ERVIN LASZLO

Books Reviewed

Science and the Akashic Field (2004)

Science and the Reenchantment of the Cosmos (2006)

Ervin Laszlo, born 1932 in Budapest, Hungary, is a philosopher of science, systems theorist, integral theorist, and classical pianist. He has published about 75 books and over 400 papers, and is editor of *World Futures: The Journal of General Evolution*. In 1993, in response to his experience with the *Club of Rome*, he founded the *Club of Budapest* to, in his words, center attention on the evolution of human values and consciousness as the crucial factors in changing course — from a race toward degradation, polarization, and disaster to a rethinking of values and priorities so as to navigate today's transformation in the direction of humanism, ethics, and global sustainability.

His book *Science and the Akashic Field: An Integral Theory of Everything* posits a field of information as the substance of the cosmos. Using the Sanskrit and Vedic term for space, *Akasha*, he calls this information field the *Akashic field* or *A-field*. He asserts that the quantum vacuum is the fundamental energy and information-carrying field that informs not just the current universe, but all universes past and present, that is, the *Metaverse*. Laszlo describes how such an informational field can explain why our universe is so exquisitely fine-tuned as to form galaxies and conscious life forms; and why evolution is an informed, not random, process. He believes that the hypothesis solves several problems that emerge from quantum physics, especially nonlocality and quantum entanglement. He also sees his hypothesis as solving the perennial disputes between science and religion. Recent works, not reviewed here, include *The Chaos Point: The World at the Crossroads*, Hampton Roads, 2006; *You Can Change the World: The Global Citizen's*

Handbook for Living on Planet Earth: A Report of the Club of Budapest, Select Books, 2003; *The Connectivity Hypothesis: Foundations of an Integral Science of Quantum, Cosmos, Life, and Consciousness*, State University of New York Press, 2003; *Evolution: The General Theory*, Hampton Press, 1996; *The Whispering Pond: A Personal Guide to the Emerging Vision of Science*, Element Books, Ltd., 1996; *The Systems View of the World: A Holistic Vision for Our Time*, Hampton Press, 1996.

Published by Sirius-C Media Galaxy LLC, 2011

Ervin Laszlo

Science and the Akashic Field

An Integral Theory of Everything
Rochester: Inner Traditions, 2004

This is after the books of Paracelsus, Mesmer, Reichenbach, Reich, Burr, Lakhovsky, Capra, Emoto, Hunt and Sheldrake the most important book I have read about the integration of the energy paradigm into Western science. Deepak Chopra, M.D. wrote about this book:

Deepak Chopra

The most brilliant, comprehensive, and intellectually satisfying integral theory of everything that I have ever read.

The author introduces the book with the following elucidation that I think is worth to be quoted in its integrality:

Ervin Laszlo

Akasha (â · kâ · sha) is a Sanskrit world meaning 'ether': all-pervasive space. Originally signifying 'radiation' or 'brilliance' in Indian philosophy akasha was considered the first and most fundamental of the five elements - the others being vata (air), agni (fire), ap (water), and prithivi (earth). Akasha embraces the properties of all five elements: it is the womb from which everything we perceive with our senses has emerged and into which everything will ultimately re-descend. The Akashic Record (also called The Akashic Chronicle) is the enduring record of all that happens, and has ever happened, in space and time.

There is something like a dialectic movement to be observed in the scientific evolution of humanity. There was first a high development of single individuals that today we call sages, who knew that all life is unity, that essential oneness is the most important feature of our cosmos, and thus that all is somehow interrelated. Then there was a phase of dissection, where the unitary and holistic thinking became split up, and where science that today we qualify as Cartesian was looking at the parts rather than the whole, and accordingly observed a cosmos that consisted of single elements without relationship to each other. That phase was rather short, some three hundred years at best, and the turn of events occurred in the lifetime and under the physics pulpit of Albert Einstein.

It was as if the Newtonian universe which created classical mechanics was going to pieces, virtually with every day, with the discovery of more and more correlations between phenomena that formerly had been considered as separated, and that were relegated to distinct scientific disciplines. And there was something like a turning point that we could fixate with the formulation of Heisenberg's uncertainty principle. While Einstein had helped to set this whole scientific revolution in motion through his early observation that a particle can be particle at times and wave at other times, and that its wave-like state collapses under observation, and thus under the impact of human consciousness, he resisted to admit that after all his relativity theory was not the last word to be said about quantum physics.

The paradoxes accumulated. In the meantime we have got to a point where holistic thinkers such as Laszlo, who see the big picture, got to summarize that the split between the matter-universe and the meta-universe is already bridged over by the fact that today we can show that the coherence factor that links all together in the cosmos is what Laszlo calls the *information field*, *Akashic field* or *A-field*, which has been called the *zero-point field* by others, and which Harold Saxton Burr has anticipated with his concept of the *L-field*. However, it has to be seen that Burr was still thinking in terms of electromagnetic fields, which is not entirely correct. When we are on the quantum level, we are beyond electromagnetic functionality, which exists prior to electromagnetic phenomena! The earlier research done on the cosmic life energy by Paracelsus, Swedenborg, Mesmer, Reichenbach, Reich and Lakhovsky was explicitly registering that the cosmic energy also is a cosmic *memory surface*, which is an aspect that was not per se apparent in the earlier research. Actually Laszlo and others have considered the esoteric notion of the *Akashic records*, which was given substantial weight with the readings of Edgar Cayce together with the notion of one single most potent agent, creator energy or creator principle, which, seen together gives with almost striking logic the notion of an energy-memory-field. Ervin Laszlo writes:

Ervin Laszlo

Science and the Akashic Field is a nontechnical introduction to the informed universe, cornerstone of a scientific theory that will grow into a genuine theory of everything. It describes the origins and the essential elements of this theory and explores why and how it is surfacing in quantum physics and cosmology, in the biological sciences, and in the new field of consciousness research. In highlights the theory's crucial feature: the revolutionary discovery that at the roots of reality there is an interconnecting, information-conserving and information-conveying cosmic field. For thousands of years, mystics and seers, sages and philosophers maintained that there is such a field; in the East they called it the Akashic Field. But the majority of Western scientists considered it a myth. Today, at the new horizons opened by the latest scientific discoveries, this field is being rediscovered. The effects of the Akashic Field are not limited to the physical world: the A-field (as we shall call it) informs all living things - the entire web of life. It also informs our consciousness./3

I won't comment here in this review on the first three chapters that are more or less a summary of the quantum paradoxes and that is, while undoubtedly written in a very elegant and comprehensive manner, of a rather descriptive nature. I will thus jump directly to Chapter Four 'Searching for the Memory of the Universe' where the author writes that our present science paradigm is not large enough to encompass the new notion of the information field and therefore has to be expanded. He notes four points that I shall quote here as they are really mark stones for further research on the matter:

Ervin Laszlo

• The universe as a whole manifests fine-tuned correlations that defy commonsense explanation.

• Astonishingly close correlations exist on the level of the quantum: every particle that has ever occupied the same quantum state as another particle remains correlated with it in a mysterious, nonenergetic way.

• Post-Darwinian evolutionary theory and quantum biology discover similarly puzzling correlation within the organism and between the organism and its milieu.

• The correlations that come to light in the farther reaches of consciousness research are just as strange: they are in the form of 'transpersonal connections' between the consciousness of one person and the mind and body of another./45

• Let's not forget to mention *Nikola Tesla* here, who was one of the early pioneers and who said, as reported by Laszlo, that the curvature of space, which was Einstein's explanation of the interconnectedness of quantum phenomena, as for example the slowing down of clocks and the shrinking of yardsticks near the speed of light, was not the answer./46

Published by Sirius-C Media Galaxy LLC, 2011

Contrary to Capra who argues more on the line of Einstein, following the non-friction reasoning of the Michelson-Morley experiments, and who was not outspoken for the existence of the ether as a valid scientific notion prior to the zero-point field, Laszlo affirmed it as an 'invisible energy field' called the luminiferous ether in his background brief 'The Quantum Vacuum' (p. 47) and showed the scientific development of it until today's belief that it is a 'physically real cosmic plenum' (p. 50). Let me quote the question before I am going to follow Laszlo's reasoning on how the vacuum can be an information field:

Ervin Laszlo

The quantum vacuum, it appears, transports light, energy, pressure, and sound. Could it have a further property by means of which it correlates separate and possibly distant events? Could it create the correlations that make for the amazing coherence of the quantum, of the organism, of consciousness - and of the whole universe? The vacuum could indeed have such a property. It could be not just a superdense sea of energy, but also a sea of information./50

And let's not forget the research of Apollo astronaut Edgar Mitchell, who basically came to this conclusion, as an answer to Laszlo's question. Laszlo reports:

Ervin Laszlo

According to Mitchell, information is part of the very substance of the universe. It is one part of the 'dyad' of which the other part is energy. Information is present everywhere, and has been present since the birth of the universe. The quantum vacuum, Mitchell said, is the holographic information mechanism that records the historical experience of matter./51

The scientific explanation of how the quantum vacuum impacts upon the historical experience of matter, Laszlo reports, was explained by a theory presented by Russian researchers and that is known as the *torsion-wave* theory. According to this theory, torsion waves, which thus must be thought of as information rays, link the universe at a group speed of the order of 109 c - one billion times the speed of light! The information aspect of the waves was explained, according to Laszlo by a Hungarian researcher as related to the spin of the particle, which results in a magnetic effect. The magnetic impulse becomes registered in the vacuum in the form of minute vortices. Laszlo writes:

> **Ervin Laszlo**
>
> These minute spinning structures travel through the vacuum, and they interact with each other. When two or more of these torsion waves meet, they form an interference pattern that integrates the strands of information on the particles that created them. This interference pattern carries information on the entire ensemble of the particles./52

As this kind of connectedness may be difficult to comprehend, Laszlo compares it with the sea. As the water of the sea interconnects all vessels, fish, and other objects in the water, and as the waves of water impact upon the motion of ships, these objects all being 'in-formed' by the motion, so do the torsion waves modulate all things in the cosmos, by creating complex patterns. Practically, we can deduce the location, speed and even the tonnage of vessels by analyzing the resulting wave-interference patterns.

And here is a couple of sentences I will note very carefully for they fully confirm Masaru Emoto's allegations on the memory surface of water, and generally his rather controversial water research. Laszlo writes in a parenthesis:

Published by Sirius-C Media Galaxy LLC, 2011

Ervin Laszlo

Water has a remarkable capacity to register and conserve information, as indicated by, among other things, homeopathic remedies that remain effective even when not a single molecule of the original substance remains in a dilution./53

The next very important clarification regards the often debated question if the quantum vacuum is really frictionless; many question that a frictionless vacuum could exist at all. Yet, Laszlo points to the recognized superfluidity of supercooled helium (2.17 Kelvin), which is a vacuum that according to John Wheeler's calculations has an energy density of 1094 erg per cubic centimeter, which is more than the energy associated with all the matter particles in the universe, when measured moving at the speed of the light. Laszlo concludes:

Ervin Laszlo

The fact is that the vacuum is both superfluid and superdense - much like helium near the absolute zero of temperature. This is a mind-boggling combination, for how can something be denser than anything else and at the same time more fluid than anything else? The vacuum, just like supercooled helium, may be a mind-boggling medium, but it is not a supernatural one./54

Now, let us look what information does to the vacuum. Laszlo speaks of a *ground state*, which is when no information flow is registered. Now, when vortices excite the vacuum, what happens is that interference patterns are created which contain the actual information. Laszlo concludes:

Ervin Laszlo

As the vortices of individual things merge, the information they carry is not overwritten, for the waves superpose one on the other. And the superposed waves are in a sense / everywhere throughout the vacuum. This, too, is a natural phenomenon: it is familiar in the form of holograms.

> In a holographic recording - created by the interference pattern of two light beams -there is no one-to-one correspondence between points on the surface of the object that is recorded and points in the recording itself. Holograms carry information in a distributed form, so all the information that makes up a hologram is present in every part of it./54-55

I would like to mention here the excellent study by Michael Talbot, *The Holographic Universe (1992)*, which I equally reviewed, and that gains a much greater importance after these revelations on the actual nature of holograms, and how they are created. Laszlo ends the chapter with the lucid statement:

Ervin Laszlo

The quantum vacuum generates the holographic field that is the memory of the universe./55

To keep this book review in reasonable boundaries, I am going to jump the 5th chapter about the cosmic fables, the paradoxes that actually triggered the thinking process from about the time when Einstein was sitting down to write his memoirs. Now, the 6th Chapter is really explosive again, as it deals in detail with the *A-Field Effect*. The author writes:

Ervin Laszlo

The A-field conveys information, and this information, subtle as it is, has a notable effect: it makes for correlation and creates coherence. This 'in-forming' of everything by everything else is universal, but it is not universally the same. Universal information does not mean uniform information. The A-field conveys the most direct, intense, and therefore evident information between things that are closely similar to one another (i.e., that are 'isomorphic' - have the same basic form). This is because the A-field information is carried by superposed vacuum wave-interference patterns that are equivalent to holograms. We know that in a hologram every element meshes with isomorphic elements: with those that are similar to it. Scientists call such meshing 'conjugation' - a holographic pattern is conjugate with similar patterns in any assortment of patterns, however vast.

Practical experience bears this out. Using the conjugate pattern as the 'key', we can pick out any single pattern in the complex wave pattern of a hologram. We need merely to insert the given wave pattern into the welter of patterns in the hologram and it attaches to patterns that are conjugate with it. This is similar to the phenomenon of resonance. Tuning forks and strings on musical instruments resonate with other forks and strings that are tuned to the same frequency, or to entire octaves higher or lower than their frequency, but not with forks and strings tuned to different frequencies./107

Here is clearly the answer that was left open in Sheldrake's *A New Science of Life (1995)* that I criticized exactly because the author never considered the cosmic energy field, or quantum field, and the newest research on it, which namely, as this quote by Laszlo clearly shows, can satisfactorily explain the phenomenon of morphogenetic resonance without needing any specific terminology such as Sheldrake's 'morphogenetic germs', which after all are a pure non-sense without having an information field connected to them that registers and stores the information. In my view, research on morphogenesis makes only sense by broadening research on the A-field, not by expanding in endless dead hypotheses that, because they are operating in a non-living universe (namely, the head of Rupert Sheldrake), are condemned to disappear before they are even written on paper.

And by the way, the simple term 'in-formation', put in the way Laszlo writes it, says it all. It says namely that in-formation forms and goes into things to change them, and this is exactly what morphogenesis says. When mouse A yesterday went successfully out of the maze in Los Angeles, and mouse B does that today in New York, then mouse A has given in-formation to mouse B that was transmitted through the A-field, and not through a morphogenetic resonance that operates without energy, because such thing cannot be in a universe that is pure energy and where nothing can exist that is not touched in one or the other way by the cosmic energy field (that for Mr. Sheldrake is part of a 'vitalistic' theory and thus has no reality per se, and here's where he deludes into paranoia, as so many materialists who, in all their rampant mysticism think they were rational; *they* are the mystics, not those they label *vitalists*). I think Sheldrake sees energy only as kinetic energy and not in its quality as an information carrier, and thus as an all-pervasive energy, and this may explain why he refuses to use the term energy for explaining morphogenetic information transmission.

Regarding evolution, I find it important that Laszlo states that while the universe does not have a definite direction, there is evolution toward growing structure and complexity. And further:

> **Ervin Laszlo**
>
> The evolution of the Metaverse is cyclic but not repetitive. One universe informs the other; there is progress from universe to universe. Thus, each universe is more evolved than the one before./131

Another important information provided by the author is the fact that evolution on planet earth cannot be explained with a simplistic early Darwinian theory of *chance mutations*, as so many of the Cartesian scientists believe it to be. Laszlo writes:

> **Ervin Laszlo**
>
> The evolution of life on Earth did not rely on chance mutations, nor did it require the physical importation of organisms or proto-organisms from elsewhere in the solar system, as the 'biological seeding' theories of the origins of life suggest. Instead, the chemical soup out of which the first proto-organisms arose was

informed by the A-field-conveyed traces of extraterrestrial life. Life on Earth was not biologically, but rather informationally seeded./136

Laszlo, as a growing number of scientists, among them Fritjof Capra, clearly contradicts the early Darwinian theory (while he probably still accepts the newer post-Darwinian theory), and interestingly the hypothesis of alien seeding as a creational myth can be found, since long, in esoteric and religious writings, channeled messages and, else, in the writings of enlightened minds such as Terence McKenna, who contended to have received this information from psychedelic mushrooms.

The good thing about Laszlo is that he takes risks for what he says, much to the contrary of many of his colleagues, and here I especially found Sheldrake to be disturbing in his refraining from definite answers that namely involve the risk as a scientist to be considered 'controversial' if not queer. Now, regarding the big question what is reality, a question so big that all religions tackle it, Laszlo gives a stupendously clear, and straightforward answer:

Ervin Laszlo

The answer to this age-old question is now relatively straightforward. The primary reality is the quantum vacuum, the energy- and information-filled field plenum that underlies our universe, and all universes in the Metaverse./140

Now, the answer how in detail reality comes about is given by Laszlo a little further down: its through in-forming all connected personal realities that are the outcome of individual holograms. He writes:

Ervin Laszlo

All we experience in our lifetime - all our perceptions, feelings, and thought processes - have cerebral functions associated with them. These functions have wave-forms equivalents, since our brain, like other things in space and time, creates information-carrying vortices - it 'makes waves'. The waves propagate in the vacuum and interfere with the waves created by the bodies and brains of other people, giving rise to complex holograms. Generations after generations of humans have left their holographic traces in the A-field. These individual holograms integrate in a super-hologram, which is the encompassing hologram of a tribe, community, or culture. The collective holograms interface and integrate in turn with the super-superhologram of all people. This is the collective information pool of humankind./150

I shall leave it here with this review. This really is a book of answers. There is much more in the book than what I could review without blowing this up to unreasonable dimensions. I think my review demonstrates, even without having read the actual book, that Ervin Laszlo is one of our most important science thinkers, and I am surely not the only one who says this.

Published by Sirius-C Media Galaxy LLC, 2011

Ervin Laszlo

Science and the Reenchantment of the Cosmos

The Rise of the Integral Vision of Reality
Rochester: Inner Traditions, 2006

This book is a sampler, not completely authored, but edited by Ervin Laszlo. But that's surely not a disadvantage. This book starts in its introduction with a statement that really made me happy. I have never believed in the big bang theory, as this theory is not in accordance with the principle of smooth continuity that can be observed in all nature. Okay, the birth of a star is a different matter, a quite violent process actually, the death of a star, too. So I think that this big bang thing was modeled after the birth of stars, but I do honestly believe that there was never a beginning of life, and that there will never be an end. We know perhaps more about the future of our galaxies, we know about the anticipating folding back of the universe upon itself, which can be seen both as end and a new beginning, both as death and as a new birth. For in nature, nothing ever dies without something else becoming manifest. So with this natural principle in mind, there can honestly not be a single timelined event called big bang. It's a typical error of modern-day scientists, and betrays their linear thinking fixation. I know this since the 1960s when as an adolescent I heard of the big bang theory for the first time on TV. I was just sixteen at that time, and remember I told my mother that in religion it was the same as with the so-called big bang. She asked how, and I explained:

– They say that we are born in this life and then we die, and after that, nothing. That cannot be.

– Why, I don't understand what you mean.

– It cannot be because there must have been something before, and there must be something thereafter.

– Why do you think that?

– It's logical, at least for me. They are not logical, the teachers, they don't know anything. One day we will know it, I am sure there was something before, and there will be something after we die. It cannot be as they say, it doesn't make sense.

My mother said she had always believed in reincarnation but that they had thrown her out of the religion class in school because she had questioned her stupid theories. I said, you see, now this begins to make sense, we reincarnate you say, so what about before we were born? Is it not logical that, then, we also had some former life? My mother nodded and said I was kind of smart, and sometimes stubborn with my inquisitiveness, and that that was probably the reason I had such a hard time in school. Well, anyway, now I am reading exactly what I intuited as a child, in a book written by somebody who should know it:

Ervin Laszlo

The widespread idea - that all there is in the universe is matter, and that all matter was created in the Big Bang and will disappear in a Big Crunch - is a colossal mistake. And the belief that when we know how matter behaves we know everything - a belief shared by classical physics and Marxist theory - is but sophistry. Such views have been definitely superseded. The universe is more amazing than classical scientists, engineers, and Marxists held possible. And the connectedness / and oneness of the universe is deeper and more thorough than even writers of science fiction could envisage./1-2

And this is, put in wonderfully simple words, quite what I had intuited all through my childhood and youth. I knew our physics teacher was wrong; that is why he hated me, and called me a *daydreamer*. I was getting more knowledge about the universe through daydreaming than he with his endless calculations. I knew it was wrong to split it all off in various disciplines, and to consider elements and single pieces of knowledge, where there was *one unity*. I knew that always, and I am pretty sure most natural children did and do, only that if I'd question now some of my friends, I would probably hear they had forgotten about all this. Anyway, now I read:

Ervin Laszlo

A cosmos that is connected, coherent, and whole recalls an ancient notion that was present in the tradition of every civilization; it is an enchanted cosmos./2

Enchantment, I would like to tell Mr. Laszlo, is an expression so far removed from the mentality that reigned in a German *Gymnasium* back in the 1960s, that it sounds like a joke, actually. And I am sure he knows that, while I know that Hungary was always more emotionally intelligent than Germany, by far more. Where are the greatest pianists of the world being born? A part in Hungary, and the other part in the Ukraine. That's simply so, accept it or not. And even today, Laszlo is really an avatar, a great senior pioneer on our way back to *real life*, and away from fake life as it was conceived by Cartesian stupidity. Yes, I think that even today it's novelty to talk about the enchantment of the cosmos, as we are not yet firmly rooted in a holistic science paradigm; we are still exploring the contours of such a new and yet so old concept. And that is why this book is important. Laszlo continues:

Ervin Laszlo

The reenchantment of the cosmos as a coherent, integral whole comes from the latest discoveries in the natural sciences, but the basic concept itself is not new; indeed, it is as old as civilization. In ages past the connectedness and wholeness of the world was known to medicine men, priests, and shamans, to seers and sages, and to all people who had the courage to look beyond their nose and stay open to what they saw. Theirs, however, was the insight that comes from mystical, religious, or aesthetic experience and was private and unverifiable - even if it appeared certain beyond doubt. Now, in the first decade of the twenty-first century, innovative scientists at the frontiers of science are rediscovering the integral nature of reality. They lift the private experiences that speak to it from the domain of unverifiable intuition into the realm of interpersonally verifiable public knowledge./2

I would say a mediocre scientist needs calculations for hiding his ignorance; a brilliant scientist knows when it's time to put his tools in the drawer and contemplate the whole of his scientific insights, in a state of meditative, and introspective intelligence. Then, he or she will really grasp the essence, that was put in these words by Laszlo:

Ervin Laszlo

We can live up to our potentials as conscious beings: we can come to know the reenchanted cosmos. Not only is this not impossible, it is not even particularly difficult. Beyond the complex deductions and abstruse mathematics of the new sciences, the basic concept of a coherent, connected, and integral universe is simple and meaningful: indeed, it is beautiful./3

This book consists of three parts. The first two parts are written by Laszlo himself, the third part contains contributions by other authors that I will list further down. The first part, *The Reenchantment of the Cosmos* is a summary of what Laszlo wrote earlier in *Science and the Akashic Field*. While this is so,

Published by Sirius-C Media Galaxy LLC, 2011

Laszlo somehow explained things still more comprehensively in this summary than I found it expressed in his earlier book, which is why I am going to put some quotes here that, while they say nothing new compared to the earlier book, are simply beautifully expressed. It is actually an extension of the field research, as it deals with biology, not physics, and with the human body. I hold these statements for particularly revealing and important for our future as human beings. Laszlo writes:

Ervin Laszlo

The vital functions of the body are governed by constant, quasi-instant and multidimensional correlations. Simple collisions among neighboring molecules - mere billiard-ball push-impact relations - do not suffice. They are complemented by a network that correlates all parts of the system, even those that are distant from one another. Rare molecules, for example, are seldom next to each other, yet they find each / other throughout the organism. This is important for the organism needs to react to stresses and strains as a whole, mobilizing all its resources wherever they are located. There would not be time for an integrated response to occur by a random process of jiggling and mixing; the molecules need to locate and respond to each other specifically, whether they are proximal or distant. (...) The body's high level of internal coherence makes possible a high level of sensitivity to the external world. In the insect world a few pheromones in the air are sufficient to attract males to prospective mates many miles away. In the human being the eye can detect single photons falling on the retina, and the ear can detect the motion of single air molecules. The mammalian body responds to extremely low frequency electromagnetic radiation, and to magnetic fields so weak that only the most sophisticated instruments can register them. Such sensitivity is only possible when a large number of molecules are coherently linked among themselves./8-9

Very important in this chapter is also Laszlo's fundamental criticism of Darwinism, and here I would like to remind that Fritjof Capra, too, has put very substantial arguments on the table that disprove most of the Darwinian assumptions about random mutations. Laszlo writes:

> **Ervin Laszlo**
>
> Living organisms are so finely tuned to their milieu that any mutation of their genome resulting from random / alterations will almost certainly reduce rather than enhance the viability of a species. Random mutations would end up by impairing fitness to the point where the species could no longer survive.
>
> However, the biosphere is populated by a vast number of complex species, the result of a long series of successful genetic mutations. This indicates that mutations in the genome are not always piecemeal and random, but are sometimes massive and systemic. If they are to be successful, the mutating elements of the genome must be highly coordinated among themselves, and must likewise be in harmony with the conditions in which the species finds itself. This suggests that the mutating genome is not fully isolated from the phenome and the environment in which the phenome finds itself. But to claim this is heresy for Darwinism, even in its current form known as the 'synthetic theory'./15/16

Under 'The Coherence of the Human Mind and the Universe', Laszlo then shows staggering scientific evidence of psychic phenomena, and I would like to point the researcher here also to the interesting studies by Dean Radin and Michael Talbot that I equally reviewed. The first feat to mention here is the synchronization of brain waves that was observed in groups during meditation. Laszlo writes:

> **Ervin Laszlo**
>
> The experiments show that as people enter an altered state of consciousness - in deep meditation or prayer - the electrical activity of the left and right frontal hemispheres of their brain be-

comes synchronized. Still more remarkable, the electroencephalograph (EEG) patterns of the left and right brain hemispheres of an entire group of persons can become synchronized with one another. In repeated tests up to twelve meditators achieved a 50 to 70 percent synchronization of their EEG waves while sitting in deep meditation in complete silence, with closed eyes and no sensory contact with each other./18

Separative Western culture has never really fostered coherence in people and between people, but my experience is that, for example, Japanese culture does very much stress, and positively value, the fact that people in a group act in some coherent, organized, and mutually supportive way - not just each for himself or herself. I have vividly seen this happening when in the train from the railway station to downtown Tokyo. There was a moment about halfway the distance that one person in the compartment was falling asleep. It was very visible, as the person had their head just hanging down, in a very carefree yet relaxed position, and I thought if this man had a stiff neck as so many Westerners, his head could not hang down so deeply as it did. And to my great surprise, about five minutes later all other Japanese were sleeping. I thought being in a movie. And when about half an hour later one of the people woke up, about five minutes later all were awake again. Now you have to see that these people were not a group, were not people who knew each other, and had been randomly put, by prior reservations, to sit with each other, or close to each other, and that compartment. I also wondered why my brain was not affected and I did not feel sleepy, while the whole compartment was asleep. And here we read:

Ervin Laszlo

A growing storehouse of evidence indicates that when the brain functions coherently, consciousness is not limited to the signals conveyed by the senses. This is a surprise to modern people who view extra- or non-sensory perception with skepticism, but it is not surprising for other cultures. Traditional tribes knew and actively used some form of extrasensory perception in their daily life. Shamans and medicine men could induce the altered state of consciousness where spontaneous information-transmission becomes possible, and their spiritual powers appear to have been a consequence of this state./19

When I apply the results of this research to Japanese people, I would have to conclude that, because they are more coherent in their relatedness, must have higher extrasensory perception abilities. And there was another information I picked up in Japan. It was about earthquakes. I had heard that earthquakes are very frequent in some parts of Japan but that there is hardly ever any damage, let alone human suffering as a result. I asked why. I got to hear that first of all in these regions houses were built from very light material, virtually paper, carton and wood, and do not contain heavy objects, nor lamps hung at the ceiling, and that besides the people were very well organized in their community spirit. That they also had a strong intuition, sensing the quake coming, and doing all preparations needed, always in joint-effort, so that all major damage was avoided. Whereupon I tried to imagine how this would be like in a Western country …

I shall leave it here with my book review, as it's already a long essay and I will not discuss the contributions here, which are interesting to read, not only because they are written by highly qualified authors, but because they all take reference to Laszlo's books and this is very valuable because it shows how Laszlo's research is received by different people, which is something very precious as it deepens your understanding of the author. We all have a bias, but when you see that a scientist and author receives a lot of respect from a lot of different people, then that person must do something that somehow has a strong impact on our lives, and our world. That would hardly happen to someone who just has some fancy ideas. And I think as

Published by Sirius-C Media Galaxy LLC, 2011

Laszlo is controversial in the established science world, this is a very good way to solidify his stance as a science and systems pioneer.

I think it is beyond doubt that this book and *Science and the Akashic Field* are very important scientific contributions that an intelligent and aware reader simply cannot miss out on. Besides, I am not competent enough to really give a judgment here, as I am not a scientist. I can only say these books have greatly enriched me.

BRONISLAW MALINOWSKI

I discovered the writings of Bronislaw Malinowski and of Margaret Mead back in 1985, in the framework of my larger research on child sexuality.[7] As early as in 1929, Malinowski published his report on the sexual life of the Trobriands in which he draws the reader's attention to the sexual life of children and adolescents.[8] Malinowski observed, not without surprise, high sexual permissiveness toward children's free sexual play. More generally, he noted the total absence of a morality that

condemns sexuality in children. Instead, he observed, children engage in free sexual play from early age.[9] Initiatory rites, Malinowski found, were absent with the Trobriands since children were initiated from about three years onwards, generally by older children, in all forms of sexual play. This play is completely non-violent and includes, with the older children, complete coitus.

7 See for example Susanne Cho, *Kindheit und Sexualität im Wandel der Kulturgeschichte, Eine Studie zur Bedeutung der kindlichen Sexualität unter besonderer Berücksichtigung des 17. und 20. Jahrhunderts*, Doctoral Thesis, Zurich, 1983; Larry L. & Joan M. Constantine, *Treasures of the Island: Children in Alternative Lifestyles*, Beverly Hills: Sage Publications, 1976 and *Where are the Kids? Children in Alternative Life-Styles*, in: Libby & Whitehurst (ed.), Marriage and Alternatives, Glenview: Scott Foresman, 1977.

8 See B. Malinowski, *The Sexual Life of Savages in North West Melanesia*, New York: Halycon House, 1929 and *Sex and Repression in Savage Society*, 1927, Reprint Chicago: University of Chicago Press, 1985.

9 Id., p. 76.

The most interesting finding for Malinowski was that in this culture violence was as good as non-existing and that there were equally as good as no sexual dysfunctions. Trobriands were found to be almost ideal marriage partners and divorce was a rare exception. Violent crimes were non-existent and incest strongly tabooed and inhibited by social norms.

Other researchers found similar phenomena with the *Muria* tribe in South India where children stay until their maturity in so-called *ghotuls* where they live their sexuality freely and in utter promiscuity. Older children initiate younger ones progressively into sexual play. These researchers found that after a phase of total promiscuity, the children, from the age of sexual maturity, form strong bonds and partnerships that are based not on sexual attraction, but on love. They further found that these first steady relationships formed the basis for later marriages that, regularly, last life-long.[10]

Some researchers and sociologists pretend that these findings had no significant meaning for our own culture since they could not be generalized. However, such arguments assume that man, depending on cultural conditioning, was basically different from one culture to the other. This is questionable, for the biological foundations are with all human beings the same, regardless of cultural or social conditioning. If all anthropological or psychological insights were valid only in a given culture, how could psychoanalysis which was founded by Sigmund Freud in Austria be successfully applied in the United States or even in India or South America?

One cannot simply disregard the extensive field studies of highly qualified anthropologists such as Malinowski or Margaret Mead or wipe them away with moralistic fake arguments, as it now seems to be the trend especially

10 V. Elwin, *The Muria and their Ghotul*, Bombay: Oxford University Press, 1947; Richard L. Currier, *Juvenile Sexuality in Global Perspective*, in : *Children & Sex, New Findings, New Perspectives*, ed. by Larry L. Constantine & Floyd M. Martinson, Boston: Little, Brown & Company, 1981, pp. 9 ff.

in the United States. Political and religious fundamentalism has many faces and often goes subtle ways in order to vacuum-clean truths that are
against their particular ideology. Interestingly,
neither Bronislaw Malinowski nor Margaret Mead
have found *pedophilia* present in Melanesia's Tro-
briand culture where children enjoy the utmost of
emotional and sexual freedom. In fact, typically, children in this culture are
sexually active with peers, and not with adults.

In other tribal cultures, a bit around the world, *pederasty* practiced with pu-
bescent boys has a quite limited function and seems to assume overall a
temporary initiatory function for the boys to be inserted in the male group.
It mainly serves to accompany the boys' come to age.

Published by Sirius-C Media Galaxy LLC, 2011

LYNNE MCTAGGART

The Field

The Quest for the Secret Force of the Universe
New York: HarperCollins, 2002

The Field by Lynne McTaggart is the book I always wanted to read because I always wanted to write it. I always wanted to write a study that proves that all what myopic Western science excludes, exists. And here is this book. It's a thriller, and the author is able to convey the complex material in understandable terms. It is obvious that she understands what she writes about, and some of the quantum physics stuff really is not easy to grasp.

The author argues from the premise that all in our universe is interconnected, that nothing is isolated, or, as scientists say, that all is *entangled*. Now, when you put up such a point of departure, a lot of consequences flow out from this. She writes:

Lynne McTaggart

Perhaps the most essential ingredient of this interconnected universe was the living consciousness that observed it. In classical physics, the experimenter was considered a separate entity, a silent observer behind glass, attempting to understand a universe that carried on, whether he or she was observing it or not. In quantum physics, however, it was discovered, the state of all possibilities of any quantum particle collapsed into a set entity as soon as it was observed or a measurement taken. To explain these strange events, quantum physicists had postulated that a participatory relationship existed between observer and observed - these particles could only be considered as 'probably' existing in space and time until they were 'perturbed', and the act of observing and measuring them forced them into a set state - an act akin to solidifying Jell-O. This astounding observation also had shattering implications about the nature of reality. It / suggested that the consciousness of the observer brought the observed object into being./11-12

This is the most lucid and well-written explanation about the observer standpoint in modern physics I ever found in a book. This is the most revolutionary consequence quantum physics provides us with: when we observe life we *change* life. So if by observing the world, we change the world, it becomes evident that we are entangled with the world – and not isolated islands in space. Now, after a few preliminaries as the present one, McTaggart opens the ring:

Lynne McTaggart
Scientists might understand in minute detail the screws, bolts, joints and various wheels, but nothing about the force that powers the engine./12

So what is that secret force that drives all, that is the invisible engine behind all, and that animates all? It's called the *zero point field*. Is that the magic formula that brings back the ether and the orgone that have been debated away to hell? In a way, yes. McTaggart shows with convincing evidence that modern quantum-enriched science has more or less integrated the cosmic energy field as it was known since millennia. And as I had predicted it years ago, it did not do this turning of the wheel in a straightforward manner, simply because it didn't want to admit that for four hundred years it was tapping in the dark. It didn't want to say that it was a shame that the Church summoned Paracelsus in front of the Inquisition, that Mesmer was unjustly shunned and exiled and that Reich definitely needs to be rehabilitated.

Published by Sirius-C Media Galaxy LLC, 2011

Instead, as we know that physicists are elegant people, it's not surprising that, for avoiding accusations of scientific neurosis, they opened the long-awaited backdoor and let the devil in from behind. They would have avoided it, for sure, and let the old Reich roast even longer in the purgatory, but the bomb that exploded in their elegant and orderly worldview was a bit too devastating. That bomb was quantum physics. And they couldn't go on, as their medical colleagues who really think there is something like 'junk DNA', continue to affirm the universe was basically empty, a vacuum tube. McTaggart writes:

> **Lynne McTaggart**
>
> Quantum mechanics had demonstrated that there is no such thing as a vacuum, or nothingness. What we tend to think of as a sheer void if all of space were emptied of matter and energy and you examined even the space between the stars is, in subatomic terms, a hive of activity./19

McTaggart also explains that our universe is not only active 'in between' matter, but is also a 'relational' interface where everything is connected with everything, and thus where all is in relationship with each other:

> **Lynne McTaggart**
>
> What we believe to be our stable, static universe is in fact a seething maelstrom of subatomic particles fleetingly popping in and out of existence. Although Heisenberg's principle most famously refers to the uncertainty attached to measuring the physical properties of the subatomic world, it also has another meaning: that we cannot know both the energy and the lifetime of a particle, so a subatomic event occurring within a tiny time frame involves an uncertain amount of energy. Largely because of Einstein's theories and his famous equation E=mc2, relating energy to mass, all elementary particles interact with each other by exchanging energy through other quantum particles, which

are believed to appear out of nowhere, combining and annihilating each other in less than an instant.../19

One of the ways of looking at subatomic particles that physicists needed to change was to see them as isolated pieces of matter. Every time when they would look at them in that way, a paradox would happen, which led to a different way of thinking. And this has altogether changed our physics. The author notes:

> **Lynne McTaggart**
>
> As the pioneers of quantum physics peered into the very heart of matter, they were astounded by what they saw. The tiniest bits of matter weren't even matter, as we know it, not even a set something, but sometimes one thing, sometimes something quite different. And even stranger, they were often many possible things at the same time. But most significantly, these subatomic particles had no meaning in isolation, but only in relationship with everything else. At its most elemental, matter couldn't be chopped up into self-contained little units, but was completely indivisible. You could only understand the universe as a dynamic web of interconnection. Things once in contact remained always in contact through all space and all time./XV

As a result of this interconnectedness, the observer-scientist cannot be seen anymore as an isolated piece of matter either. It was only after including the observer in the experiment that paradoxes could be avoided and comprehensive results were achieved in quantum physics. In fact, the observer was excluded only about four hundred years ago from science, with the shift to the Cartesian science paradigm. The author writes:

> **Lynne McTaggart**
>
> Perhaps the most essential ingredient of this interconnected universe was the living consciousness that observed it. In classical physics, the experimenter was considered a separate entity, a

silent observer behind glass, attempting to understand a universe that carried on, whether he or she was observing it or not. In quantum physics, however, it was discovered, the state of all possibilities of any quantum particle collapsed into a set entity as soon as it was observed or a measurement taken. To explain these strange events, quantum physicists had postulated that a participatory relationship existed between observer and observed - these particles could only be considered as 'probably' existing in space and time until they were 'perturbed', and the act of observing and measuring them forced them into a set state - an act akin to solidifying Jell-O. This astounding observation also had shattering implications about the nature of reality. It / suggested that the consciousness of the observer brought the observed object into being./11-12

The immense energy that has been measured as pertaining to the Zero Point Field could represent another piece of evidence to its 'global motor' kind of function in our universe. McTaggart writes:

Lynne McTaggart
It has been calculated that the total energy of the Zero Point Field exceeds all energy in matter by the factor of 1040, or 1 followed by 40 zeros./23

It also has been found that the Zero Point Field contributes to the stability of matter and represents something like a blueprint of the whole universe:

Lynne McTaggart
You can show mathematically that electrons lose and gain energy constantly from the Zero Point Field in a dynamic equilibrium, balanced at exactly the right orbit. Electrons get their energy to keep going without slowing down because they are refueling by tapping into these fluctuations of empty space. In other words, the Zero Point Field accounts for the stability of the hydrogen atom - and, by inference, the stability of all matter./25

> If all subatomic matter in the world is interacting constantly with this ambient ground-state energy field, the subatomic waves of The Field are constantly imprinting a record of the shape of everything. As the harbinger and imprinter of all wavelengths and all frequencies, the Zero Point Field is a kind of shadow of the universe for all time, a mirror image and record of everything that ever was./26

I wish all readers of this book much joy in the discovery of a new branch of science that needs to do some educational work so that non-scientists can obtain the privilege of understanding its coded language. McTaggart has done an important educational work with this daring study, and more is surely to follow. To summarize, the book, while the author is able to lighten the tightly scientific text with anecdotes, is really not an easy read. But that is probably not her fault but has to do with the high level of abstraction in latest-date quantum physics.

Published by Sirius-C Media Galaxy LLC, 2011

FRANZ ANTON MESMER

Franz Anton Mesmer (1734-1815) was a
German physician who was perhaps the first
medical doctor who, in newer history, has
acknowledged the influence of lunar cycles
on the physical body, and generally, by doing so, was discovering the truth behind
psychosomatic medicine. Interestingly,
Mesmer wrote his doctoral thesis about the
influence of planetary energies upon the
human body. Contrary to Paracelsus' strong
focus on plants, Mesmer's scientific and

medical focus was upon humans and animals. He did not consider plants,
and as his terminology suggests, saw humans on the same level, energetically speaking, more akin to animals.

I found Mesmer's book *Le Magnétisme Animal*, in its original French version,
in 1975, when I just had entered law school. I was thus reading Mesmer before I read Freud – and this of importance because Freud later denied the
vitalistic meaning of libido, which originally is a term synonymous with
animal magnetism.

Mesmer first experimented with magnets for healing hysteria, came up with
the expression *animal magnetism* for describing the cosmic energy. Mesmer
termed it such for the simple reason to distinguish this variant of magnetic
force from those which were referred to, at that time, as mineral magnetism,
cosmic magnetism and planetary magnetism. Mesmer saw that this vital
energy only resides in the bodies of humans and animals. He chose the
word animal, and not human, because it goes back to a Latin root: *animus*.

In Latin, animus means what is 'animated' with life, what breaths, what thus belongs to the animate realm.

What Mesmer discovered here was thus the bioplasmatic energy that since long was known before him, for example by Paracelsus who called it vis vitalis. That Mesmer found this energy to be existent only in humans and animals was his particular myopic view; long before him, Paracelsus namely showed that this energy is primarily to be found in plants, and the plant realm is much older than the animal kingdom, which is why, somehow, Paracelsus was the greater scholar here.

Mesmer, first encountered the healing currents through strong magnets that he placed between himself and the patient, and later observed, to his great astonishment, that the same healing effects occurred also without the magnets. Which made him conclude that ultimately it was his own body electrics, his own bioplasmatic vibration that positively affected the emonic current of his hysteric patients.

To conclude, Mesmer thus discovered the same subtle energy that before him Paracelsus called *vis vitalis* and that Swedenborg termed *spirit energy*. Behind the divergence in terminology, these scientists observed and reported basically the same natural phenomena.

ASHLEY MONTAGU

Ashley Montagu (1905-1999), who was born Israel Ehrenberg in East London, was a brilliant scholar, psychologist and anthropologist who earned a PhD in anthropology in 1936 at Columbia University. Montagu was a long-term lecturer and chairman of the Department of Anthropology at Rutgers, and an outstanding popular writer and visiting guest lecturer at major American universities

I found Ashley Montagu's book *Touching: The Human Significance of the Skin* back in 1984, but it took me more than a decade to process this important information. The book, with its detailed research background and many references to other, similar, research on the importance of *touch*, opened my eyes as to the importance of early tactile stimulation.

Reading Montagu's book, I had found at quite the same time the research of James W. Prescott and Herbert James Campbell as well as the writings of the great French obstetricians Frederick Leboyer and Michel Odent.

Ashley Montagu

Touching

The Human Significance of the Skin
New York: Harper & Row, 1978

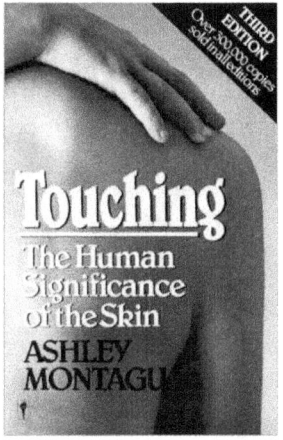

Ashley Montagu's research on touch must in fact be seen under the greater header of research into the importance of pleasure, and how the human brain relates to, and processes, pleasure. Montagu's research on rhesus, a project that went over thirty years of laboratory research, left no doubt that the human skin as a primary pleasure provider.

Ashley Montagu's research is highly interesting with regard to our understanding of tactile stimulation in early childhood. Montagu's specific focus during his research was upon the mammal mothers' licking the young. He found astonishing unity in zoologists' opinions as to the importance of this motherly licking for the survival of the offspring. He discovered that it was in the first place the perineal zone (between anus and genitals) of the young that the mother preferably and repeatedly licked. Experiments in which mammal mothers were impeded from licking this zone of the young resulted in functional disturbances or even chronic sickness of the genito-urinary tract of the young animals.

Ashley Montagu concluded from this research that the licking did not serve hygienic purposes only, but was intended to provide a tactile stimulation for the organs that were underlying the part of the skin that was licked.

From this research, important conclusions and extrapolations can be made as to the importance of human touch for peace research and human survival. It was this research that contributed to the many positive changes in the birthing experience and the improvements in early childcare, that were fed by the insights into the lifelong damages caused by early tactile deprivation as a major factor in the etiology of violence.

I report and analyze Montagu's highly interesting findings in several of my publications, showing their impact for dealing with our current problems of rampant depression, isolation of the elder, child psychosis, autism, and domestic violence. Lack of touch, and its extreme form, tactile deprivation, is namely a common factor in all these etiologies.

Published by Sirius-C Media Galaxy LLC, 2011

CANDACE B. PERT

Molecules of Emotion

The Science behind Mind-Body Medicine
New York: Scribner, 2003

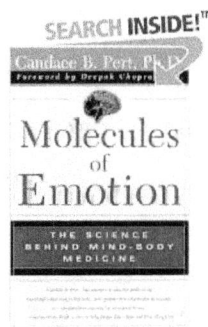

I came to know **Candace B. Pert** through the mind-boggling film *What the Bleep Do We Know (2005/2006)*. I found her presence markedly impressive and right away ordered her book. It is true that, what some people object, is quite personal in her book, and that for this reason, the book is anecdotic in some way. Anecdotic – what does that mean? It means that somebody writes about life, and not just about ivory tower ideas that spring up in a frigid mind. There is about no larger topic than *life*.

Well, this is to say that this book really reveals important novelty about life, but should this knowledge be presented in a dry statistical manner, in an emotionally detached, and cold style? I don't think so. I rather tend to think that Candace Pert very intently presented her book in that way, and let me speculate about some of the reasons. For example, the author may have wanted to establish a warm, empathic communication with the readers of her book, and she may have wanted to show the actual link between science and real life; in fact, her research was stirring up and changing the basic parameters of her life, thereby providing the best evidence ever for quantum physic's postulate that the observer and the observed are entangled, even up to daily life matters. And last not least, she may have been very well aware of the particular technical complexity of her research which, when related in a dry, statistical and cold manner, could have frightened away many readers. I think her particular mix was a real benediction for I am sure I would not have otherwise read the book until the end.

Molecules of Emotion is for these and perhaps other reasons not only an extraordinary scientific study, but it also comes with much autobiographic content. Candace Pert has the courage to reveal many details from her life as a female scientist, who is not only female, which is in her words already an obstacle to success in science, but who is also daringly original. Since the 1970s she has persisted in her vision of finding molecular evidence for the functionality of our emotions, and our sexuality, and more generally for mindbody medicine, within the boundaries of Western science.

The book, if all that additional information was taken out, would be a research paper, too thin to fill a book. And it would probably miss its goal entirely. It's this holistic and empathic approach, and needless to add that it's an artistic approach as well, that makes this book so unique. And it shows that this scientist is actually a great human. In fact, clearly not for nothing was she one of the few scientists who had the honor to be interviewed for the film *What the Bleep Do We Know* and even more extensively for its second edition, the *Rabbit Hole*. And she makes a strong point in the film. Actually Pert, together with the brilliant animations in the film, made transparent how human sexuality works, and that it is not a mechanical abstract function, that it is not, as I am saying since years and years against mainstream psychology, an instinct or *drive* as Sigmund Freud called it, but a direct outflow from our emotional predilections. To give an example, how she explains this rather complex matter in a very readable, comprehensive way, let me put this quote:

> **Candace B. Pert**
> If receptors are the first components of the molecules of emotion, then ligands are the second. The word ligand comes from the Latin ligare, 'that which binds', sharing its origin with the

word religion. Ligand is the term used for any natural or man-made substance that binds selectively to its own specific receptor on the surface of a cell. The ligand bumps onto the receptor and slips off, bumps back on, slips back off again. The ligand bumping on is what we call the binding, and in the process, the ligand transfers a message via its molecular properties to the receptor. Though a key fitting into a lock is the standard image, a more dynamic description of this process might be two voices - ligand and receptor - striking the same note and producing a vibration that rings a doorbell to open the doorway to the cell'./24

From that perspective, sexuality loses much of its myth to be a mere automatism, and it becomes obvious that it's a matter of taking options and making choices at any given moment in life. This is so because neuropeptites do not behave randomly but as a function of consciousness, as a function of conscious thinking, of intent. In addition, these new scientific insights show that sexuality is a moving dynamic thing, not a static conditioned soup that you've eaten once and that stays in your guts for the rest of your life. This means that we can change sexual conditioning, if we want to.

Candace Pert's research showed that sexuality is conscious because neuropeptites are functioning as agents of consciousness, so to say. That's about what I can say about this book. You have to read it to know all about it. This research is much too complex for being reported in a few lines. But this book truly is important, perhaps even one of the books of the century. If you question that, please do inquire about the popularity of the book, and Pert's seminars - and you will be surprised! Her popularity is amazing because for one time it's not based upon manipulation, big media propaganda, and huge fund raising from the side of multinationals to push up sales, but on the powerful and simple human personality and amazing persistence of a female scientist!

Candace Pert's project was since its humble beginnings in the 1970s very daring, as until now mainstream psychology treats emotions as 'floating parameters' that are hard to grasp by our Cartesian science paradigm. Candace Pert gives a hint how this abstruse paradigm came about in the first place:

> **Candace B. Pert**
>
> If psychological contributions to physical health and disease are viewed with suspicion, the suggestion that the soul - the literal translation of psyche - might matter is considered downright absurd. For now we are getting into the mystical realm, where scientists have been officially forbidden to tread ever since the seventeenth century. It was then that René Descartes, the philosopher and founding father of modern medicine, was forced to make a turf deal with the Pope in order to get the human bodies he needed for dissection./18

But in her own words, her vision even went beyond. She did not just want to succeed in her personal research project, but desired to help bring about this huge paradigm shift to many scientists who are currently working on it. And she wanted this paradigm shift to expand also into medical science, so that the psychosomatic unity of body and mind are definitely recognized in Western medicine. In her own words:

> **Candace B. Pert**
>
> My intention is to provide an understanding of the metaphors that express a new paradigm, metaphors that capture how inextricably united the body and the mind really are, and the role the emotions play in health and disease./17

Published by Sirius-C Media Galaxy LLC, 2011

And she explains why it is such a brain-drain and needs so much persistence to work on the leading edge of science, despite the importance of scientific progressing being made:

Candace B. Pert

Truly original, boundary-breaking ideas are rarely welcomed at first, no matter who proposes them. Protecting the prevailing paradigm, science moves slowly, because it doesn't want to make mistakes. Consequently, genuinely new and important ideas are often subjected to nitpickingly intense scrutiny, if not outright rejection and revulsion, and getting them published becomes a Sisyphean labor./19

It is known from the film *What the Bleep Do We Know* how brilliantly Pert explains her research, how she can convey complex matters in a simple comprehensive way. And this is the way she explains in her book what emotions are, or what they appear to be, contrary to mainstream psychology, under the particular angle of her research:

Candace B. Pert

When I use the term emotion, I am speaking in the broadest of terms, to include not only the familiar human experiences of anger, fear, and sadness, as well as joy, contentment, and courage, but also basic sensations such as pleasure and pain, as well as the 'drive states' studied by the experimental psychologists, such as hunger and thirst. In addition to measurable and observable emotions and states, I also refer to an / assortment of other intangible, subjective experiences that are probably unique to humans, such as spiritual inspiration, awe, bliss, and other states of consciousness that we all have experienced but that have been, up until now, physiologically explained./131-132

Generally, her emphasis both in her research and her book is upon the psychosomatic unity of the mindbody. Let me put here two more quotes to come to an end of my review:

Candace B. Pert

The body is the unconscious mind! Repressed traumas caused by overwhelming emotion can be stored in a body part, thereafter affecting our ability to feel that part or even move it./141

These recent discoveries are important for appreciating how memories are stored not only in the brain, but in a psychosomatic network extending into the body, particularly in the ubiquitous receptors between nerves and bundles of cell bodies called ganglia, which are distributed not just in and near the spinal cord, but all the way out along pathways to internal organs and the very surface of our skin. The decision about what becomes a thought rising to consciousness and what remains an undigested thought pattern buried at a deeper level in the body is mediated by the receptors. I'd say that the fact that memory is encoded or stored at the receptor level means that memory processes are emotion-driven and unconscious (but, like other receptor-mediated processes, can sometimes be made conscious)./143

To summarize, this highly readable book from an amazing scientist may scramble you up a bit inwardly, and perhaps also outwardly (through coincidences!), and this is a good thing to happen. This book is not a dry boring research report, but in the contrary reads like an adventure novel - the novel of a daring woman who has realized much in her life. She has won the hearts of many people and through touching their hearts she has been able to put new seeds in their minds. I congratulate her! Truly, her story is not one you hear often in life.

Published by Sirius-C Media Galaxy LLC, 2011

JAMES W. PRESCOTT

I discovered the writings of **James W. Prescott** in the 1980s, at a time when I was reading the books of Ashley Montagu, Frederick Leboyer, Michel Odent, Alexander Lowen, Bronislaw Malinowski and Margaret Mead. The two major articles[11] written by James W. Prescott, that I discuss in several of my own books, were coming to me like a revelation to a question I had asked since more than a decade: 'What are the roots of violence?'

Knowing from anthropological studies and from spiritual work that violence is not the natural condition for humanity, but a sort of emotional and cultural perversion that results from deep hurts early in childhood, and probably also from scars that go back to former lives, I was grateful to have found at least one conclusive research that not only analyzed our condition, but also pointed to viable solutions for creating a more peaceful society of the future.

The solutions that James W. Prescott, a developmental neuropsychologist and cross-cultural psychologist, suggests are changes in the process of child birth and our educational system, a permissive and non-violent child-rearing paradigm, social permissiveness regarding pre-marital sex and a definite legal prohibition of physical punishment of children in both the home and school together with effective government collaboration in fighting domestic violence. Regarding infant care, Prescott stresses the importance of the primary symbiosis between mother and infant during the

---

11 *Body Pleasure and the Origins of Violence, Bulletin of the Atomic Scientists, 10-20 (1975)*, and *Deprivation of Physical Affection as a Primary Process in the Development of Physical Violence, A Comparative and Cross-Cultural Perspective*, in: David G. Gil, Ed., Child Abuse and Violence, New York: Ams Press, 1979.

first 18 months of the infant, abundant tactile stimulation of infants and babies, using techniques of child massage, as well as co-sleeping between parents and small children.

It is interesting to note that the suggestions that James W. Prescott comes up with from his perspective as a peace researcher are very much in accordance with those suggested by Jean Liedloff, in her book *The Continuum Concept (1977/1986)*, from her perspective of the lifestyle of native peoples. Also, there is a striking similarity of solutions offered for the same questions by Ashley Montagu, as a result of his decades of skin research, and by the French obstetricians Michel Odent and Frederick Leboyer who have looked beyond the fence of obstetrics and into what Odent called *Primal Health*, which is a holistic concept of health and well-being.

Published by Sirius-C Media Galaxy LLC, 2011

James W. Prescott

The Origins of Love and Violence

Sensory Deprivation and the Developing Brain
Research and Prevention
DVD, 200 min.
Touch the Future, Inc., 2009

About the Author

James W. Prescott, PhD began his career researching maternal deprivation with his early life experiences in an orphanage. Later he focused on the impact sensory deprivation has on the developing brain during this sensitive-formative period from birth through adolescence. *The Origins of Love and Violence* gathers and makes available in DVD and print formats fifty years of research on early brain development and recommends two simple things most every family can do to insure that their baby develops the capacity to love (experience pleasure bonding and avoid intentional pain instead of becoming violent.

Prevention

For thousands of years we have been trying to 'talk' ourselves out of violence--a neocortical exercise-when the problem resides in our (subcortical) emotional-social-sexual brain. The Origins of Love and Violence are encoded in that developing emotional-social-sexual brain very early in life, long before most realize.

Our so called thinking (neocortical) brain developed after and is profoundly influenced by what has been programmed into the subcortical brain. In the subcortical brain, for example, pleasure is a good which attracts; pain is bad and is avoided. This equation is reversed in the neocortical brain when pleasure is equated with evil, to be avoided; and pain, suffering and deprivation are a "good", supported by a neocortical belief in "salvation". Cultural and theistic value systems invert millions of years of psychobiological evolution resulting in a war between the body and the mind.

Modern therapeutic experiences fail to recognize this basic conflict. Manipulation of the neocortical symbolic brain does not address the basic needs of the emotional-social-sexual brain which is the primary source of our dysfunctions.

Prevention by providing the nurturing experiences the developing emotional-social-sexual brain must have is the key to personal and global peace.

– James W. Prescott, PhD.

Need & Benefits

Affectionate touch, movement and play are as basic and vital to healthy development as vitamins, nutrition, sleep, exercise and language development.

Prevention involves everyone: mothers, fathers, pediatricians, nurses, childbirth educators, grandparents, pre-school educators, childcare providers, every teacher at every grade level, college and university professors, policy makers, coaches, therapists and councilors.

Emotional-social-sexual intelligence must be nurtured and developed with the same care, resources and steady dedication we now invest in reading, writing and mathematics. The sensory experiences of affectionate touch, movement, breast feeding and play are the nutrients that build a healthy emotional-social-sexual brain in the same way that vitamins, minerals, water, exercise and sunshine build a healthy body.

We all understand how sleep deprivation throws everything off, physically, emotionally and mentally. The same is true of affection-pleasure deprivation. The absence of pleasure, sensory deprivation, can be compared to chronic malnutrition or sleep deprivation. Deprive the early developing brain of the sensory experiences it needs and you stunt the development of that brain for life.

Jim's research balances the developmental equation by given equal attention, value and resources to emotional-social-sexual development as we now give to healthy bones and an

Published by Sirius-C Media Galaxy LLC, 2011

agile mind. The care and nurturing of the emotional-social-sexual system is as basic necessary as sleep and good nutrition.

Everyone can understand, benefit from and apply the insights found in this comprehensive package. Everyone can add emotional-social-sexual development to their daily diet of physical and mental nutrition and exercise.

Prescott's vast body of work makes abundantly clear emotion-social-sexual malnutrition looks like and how it retards human development lifelong.

Our goal is to awaken and develop critical appreciation and value for this grossly overlooked aspect of human development and to do so in the widest possible way.

– Michael Mendizza

DEAN RADIN

Books Reviewed

The Conscious Universe (1997)

Entangled Minds (2006)

Dean Radin is an extraordinary researcher. He was able to trigger a major paradigm shift in science, in much the same way, as Fritjof Capra did before him. Radin's research on parapsychology and psychic experiences led to a more or less widespread scientific acceptance of these phenomena, and this is really a revolution in science.

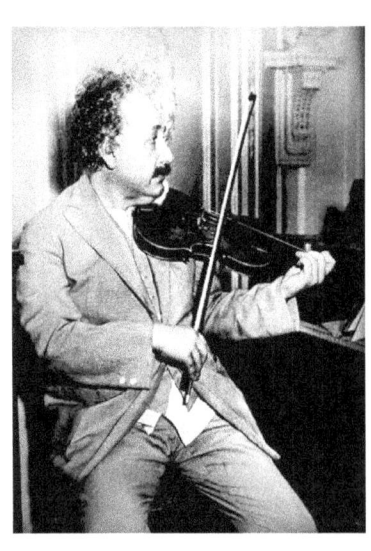

The secret of his success is a combination of talents that we know from the greatest scientists, and especially Albert Einstein; it's intuition coupled with concise logic, meticulous attention to detail and flawless methodology. In addition, Radin had the stoic mindset necessary for waiting until the time had come for society to accept the paradigm change. It was not immediate, and Radin encountered a considerable amount of resistance from both the science establishment and activist social groups who, for one reason or the other, were against his research.

Radin also has a talent for expressing himself with an ease both in speech and write that is not very often to be found with lab scientists. His face reminds Shakespeare, and his eyes express the depth of soul typically to be found with great poets. And he is a violinist, coincidentally like Albert Einstein. This man is also a great educator, and has given media presentations on his research that stand out by their clarity, and Radin's refreshing sense of humor.

Reading his books has greatly inspired me, and not only his books. Radin is also active in the online world, and he's managing an extended blog. I would like to quote a passage from it that shows that this man is a multidimensional personality, gifted with an array of talents, and interested in many subjects:

Dean Radin

I played classical violin professionally until age 25, then switched to fiddle and banjo and played in bluegrass bands for a number of years. Along the way I graduated with a degree in electrical engineering, magna cum laude and with senior honors, from the University of Massachusetts (Amherst), a masters in electrical engineering from the University of Illinois (Champaign-Urbana), and a PhD in psychology, also from the University of Illinois.

For a decade I worked at AT&T Bell Laboratories and later at GTE Laboratories on advanced telecommunications R&D, and then I held appointments at Princeton University, University of Edinburgh, University of Nevada, SRI International and Interval Research Corporation, where I was engaged in research on psychic or "psi" phenomena. At SRI I worked on what is now popularly known as the (formerly classified) psi research program condemned StarGate. In 2000 I cofounded the Boundary Institute and since 2001 I've been Senior Scientist at the Institute of Noetic Sciences. I also hold an adjunct appointment at Sonoma State University and am on the Distinguished Consulting Faculty at Saybrook Graduate School.

The majority of my professional career has focused on experimentally probing the far reaches of human consciousness, primarily poorly understood phenomena like intuition, gut feelings and psi phenomena. Very few scientists are actively engaged in research on these perennially interesting topics, and perhaps because of this unusual choice of profession I was featured in a New York Times Magazine article in 1996.

My interests in these topics were motivated partially by sheer curiosity, but also by an appreciation that these experiences are responsible for most of the greatest inventions, artistic and scientific achievements, creative insights, and religious epiphanies throughout history. Understanding this realm of human experience thus offers more than mere academic interest -- it touches upon the very best that the human intellect and spirit have had to offer. I discovered while working on these topics that I enjoy the challenge of exploring the frontiers of science, and that I am comfortable tolerating the ambiguity of not knowing the 'right answer,' which is a constant companion at the frontier.

After being engaged in the scientific investigation of such phenomena for about 25 years, I've become convinced through the laboratory evidence that some psychic experiences are genuine, that many people do have real psychic experiences (occasionally), and that most people who claim to have extremely reliable or accurate psychic abilities are delusional. This topic is exploited for entertainment purposes, and the world is full of unscrupulous individuals who falsely claim psychic abilities, so I understand why many scientists avoid this topic. Nevertheless, because the empirical evidence reveals that some psychic effects can be repeatedly observed under controlled conditions, these phenomena are profoundly important because they suggest that prevailing scientific assumptions about human capacities are seriously incomplete.

There is certainly room for scholarly debate about these topics, and I know many informed skeptics whose opinions I value. However, I've also learned that there are some who are irrationally hostile about this topic, yet they know little or nothing about it. There is no kind way to say this, but the most stubborn skeptics do not understand scientific methods or the use of statistical inference, nor do they appreciate the history, philosophy or sociology of science. Their emotional rejection of the evidence seems to be motivated by fundamentalist beliefs of the scientistic or religious kind.

Source: http://www.deanradin.com/NewWeb/bio.html

I became aware of Dean Radin's research through the movie *The Rabbit Hole Edition*, the second episode of *What the Bleep Do We Know (2006)*. And I ordered his books right away, and was amazed that this man was able to shatter the coarsest prejudice against paranormal phenomena and parapsychology. And how? He was defeating the enemy with his own weapons; he applied the purest Cartesian method of rigid trial and error, and meticulous detailed proof, and step-by-step elucidation of scientific facts, and he did

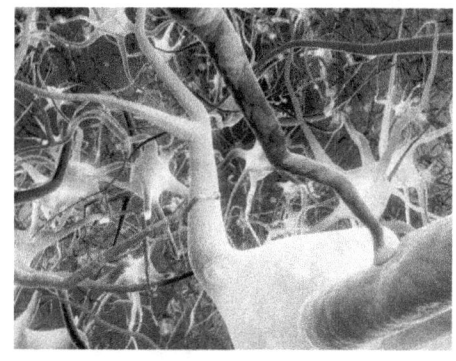

this so brilliantly that it is today simply impossible to refute his findings. In the contrary, they were corroborated by other researchers who replicated the experiments.

Well, reading him, I became aware that it was not only his rigorous research and application of the scientific method. It was also his literary genius. This man knows to write, he knows to argue, without throwing people things over their head - as so many of the scam artists do in this area of flow between charlatanism and so-called established science.

But the main thing is the strength of his vision. He set for himself the vision that parapsychology is to be defined as correct and official, and exact science. And then he started out. And he got where he wanted to get at. And credibility, yes, he had to built, and a lot of it, for becoming an authority in such a daring discipline that for decades was shunned by 'official' science and relegated to the 'unofficial' yet enlightened pulpit of esoteric freaks, geniuses, psychics, curious lawyers such as myself - and gifted children. And his brilliant methodology certainly was one of the decisive factors of his success; next to his visionary quest and outstanding communication abilities. And there we are, virtually transformed as a group, as a society, where we can observe that with every day the majority and the minority are changing roles, and it's now according to recent polls indeed the majority, at least in America, who believe that psychic phenomena are real and should be really and scientifically investigated.

And the funny thing is that the government, the military and the CIA were since long taking psi serious and were investigating it, and not with minor investments and efforts, and still, official science was denying it. It was a paradoxical situation for many years. Now, as the polls are showing that a majority of the population is convinced that psychic phenomena are real, there is also a democratic quest at stake as from a constitutional point of view, science cannot just disregard such a fact and continue to stubbornly refuse using their funding for proper research. So now, after the breakthrough, I would say that the social picture is one that makes much more sense, after all, and a lot of tensions that are not conducive to good human relations have been alleviated through the paradigm change in official Western science.

Published by Sirius-C Media Galaxy LLC, 2011

Dean Radin

The Conscious Universe

The Scientific Truth of Psychic Phenomena
San Francisco: Harper & Row, 1997

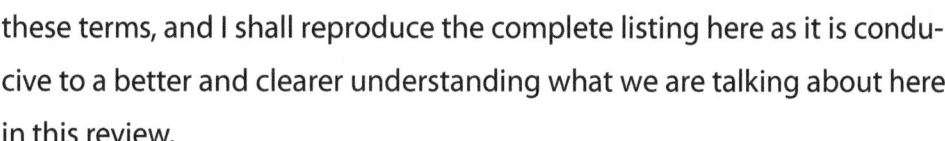

The author defines psychic phenomena not in a
vacuum, but uses the popular custom to define
these terms, and I shall reproduce the complete listing here as it is conducive to a better and clearer understanding what we are talking about here
in this review.

But for the anecdote, I may advance here that number of my personal
friends have abandoned me for nothing but the reason I am interested
since years in these phenomena, and spirituality in general, while they were
encouraging my research on emotions, on paraphilias, or on the roots of
violence. And it was always men, that is I think significant, women having a
so much easier access to the realms and qualities that are associated with
our right brain. The following definitions are to be found on pages 14 and
15 of the book.

Telepathy
Information exchanged between two or more minds, without the use of the ordinary
senses

Clairvoyance
Information received from a distance, beyond the reach of the ordinary senses. A French
term meaning 'clear-seeing'. Also called 'remote-viewing'.

Psychokinesis
Mental interaction with animate or inanimate matter. Experiments suggest that it is more
accurate to think of psychokinesis as information flowing from mind to matter, rather than

as the application of mental forces or powers. Also called 'mind-matter interaction', 'PK', and sometimes telekinesis'.

Precognition

Information perceived about future events, where the information could not be inferred by ordinary means. Variations include 'premonition', a foreboding of an unfavorable future event, and 'presentment', a sensing of a future emotion.

ESP

Extrasensory perception, a term popularized by J. B. Rhine in the 1930s. It refers to information perceived by telepathy, clairvoyance, or precognition.

Psi

A letter of the Greek alphabet (Ψ) used as a neutral term for all ESP-type and psychokinetic phenomena.

Related Phenomena

OBE

Out-of-body experience; an experience of feeling separated from the body. Usually accompanied by visual perceptions reminiscent of clairvoyance.

NDE

Near-death experience; an experience sometimes reported by those who are revived from nearly dying. Often refers to a core experience that includes feelings of peace, OBE, seeing lights, and certain other phenomena. Related to psi primarily through the OBE experience.

Reincarnation

The concept of dying and being reborn into a new life. The strongest evidence for this ancient idea comes from children, some of whom recollect verifiable details of previous lives. Related to psi by similarities to clairvoyance and telepathy.

Published by Sirius-C Media Galaxy LLC, 2011

Haunting

Recurrent phenomena reported to occur in particular locations, including sightings of apparitions, strange sounds, movement of objects, and other anomalous physical and perceptual effects. Related to psi by similarities to psychokinesis and clairvoyance.

Poltergeist

Large-scale psychokinetic phenomena previously attributed to spirits but now associated with a living person, frequently an adolescent. From the German for 'noisy spirit'.

Before going into the detailed discussion of this book, allow me a personal remark. I would like to express my deep admiration for the incredible and really tough work the author has done to realize his vision, and get parapsychology aligned with all the other sciences, as a not less exact, not less serious, and not less important science than physics, mathematics, biology and all the rest. And while this book reads very cool, as it's written with an intention focused on facts that are experimentally verified and that are coherently aligned into one or several theories, it can be verified or falsified, according to strict scientific logic and practice.

Now, reading the Postscript of the book, the reader may get an idea what the author actually went through, as a human being, as a non-conforming and novelty-oriented scientist, one of those we call the leading edge in modern science. When a project is finalized, people always sit back and look at the immense work, applauding, at least most of the time, the originator. But have they ever felt how it was to go through all that, from the first to the last moment? I personally know it, for until now, after more than twenty years of research, and several dozens of books written in three major languages, I am still in the starting holes, as nobody, absolutely nobody in the world of science has ever considered the content of my books, except for slandering me or assaulting me, or by impersonating me. But of course, here I am not talking about myself, so let's listen to what author has to say about the dark side of the moon, while in the media he's so often in the sun, in the spotlight now, as one who has greatly succeeded in his scientific career.

Dean Radin

For example, on Monday, I'm accused of blasphemy by fundamentalists, who imagine that psi threatens their faith in revealed religious doctrine. On Tuesday, I'm accused of religious cultism by militant atheists, who imagine that psi threatens their faith in revealed scientific wisdom. On Wednesday, I am stalked by paranoid schizophrenics who insist that I get the FBI to stop controlling their thoughts. On Thursday, I submit research grants that are rejected because the referees are unaware that there is any legitimate evidence for psi. On Friday, I / get a huge pile of correspondence from students requesting copies of everything I've ever written. On Saturday, I take calls from scientists who want to collaborate on research as long as I can guarantee that no one will discover their secret interest. On Sunday, I rest, and try to think of ways to get the paranoid schizophrenics to start stalking the fundamentalists instead of me./299-300

I have only one comment. It's really painful to be intelligent in an utterly stupid society, and this has little to do with our times, but can be seen over the whole course of human scientific history! So let's go in medias res and look at the little critter that makes out this rich and long book.

To begin with, I would like to stress the energy nature of those various phenomena that we use to call *paranormal*, as this is the result of my own research on the matter, that I did within my own research on emotions and bioenergy. When I speak of bioenergy here I mean the bioplasmatic energy that is also called cosmic life energy or bioenergy, and not body electrics or electromagnetism. This is also the energy that is meant and referred to in shamanism, when shamans talk about the spirits of nature. These spirits, to

Published by Sirius-C Media Galaxy LLC, 2011

be true, are energy streams that bear transcoded information, and as such they are part of the huge communication network built into living systems.

The author says that psi research does not fit in conventional theories and that it's not correct that researchers, because they face a novelty that scares them, explain what is so far unexplainable with the theory of electromagnetism.

Dean Radin

The results show that when telepathic receivers are isolated by heavy-duty electromagnetic and magnetic shielding (specially constructed for rooms with steel and copper walls), or by extreme distance, they are still able to obtain information from a sender without using the ordinary senses. So we know that telepathy doesn't work like conventional electromagnetic signaling. And yet, because the metaphor provides a powerful way of thinking about telepathy, many people still imagine that telepathy 'works' through some form of mental radio./16

I may be allowed to add that Wilhelm Reich has explained very explicitly the difference between what he called the orgone and which he held responsible not only for all life functions but also for psychic phenomena, on one hand, and electrical or electromagnetic phenomena, on the other. I have verified the matter for my essay on Wilhelm Reich and contacted the Wilhelm Reich Museum in Maine with this question[12]. The director, Ms. Mary Boyd Higgins, clearly affirmed that the books and manuscripts by Wilhelm Reich contain the proof that the orgone energy is prior to phenomena such as electricity or electromagnetism, that it well induces such phenomena, but that it is itself not explainable with any of these concepts, simply because other laws apply for it, laws that conventional science hitherto more or less completely ignores.

---

[12] www.wilhelmreichmuseum.org

What I have held against Sheldrake's theory of morphogenetic resonance in my review of his book *A New Science of Life (1995)* is that he explains morphogenetic correlations, much like Radin in the present book, with quantum teleportation that one can imagine as some sort of synchronicity in the quantum field, and that is said to operate independently of energy, as a pure information system. Radin writes:

> **Dean Radin**
>
> The link to psi is that biological systems are exquisitely sensitive to certain kinds of information. Perhaps biological systems can both send and access teleported information, in which case we would suddenly have a scientifically acceptable (but still fundamentally mysterious) way to both perceive and influence objects at a distance./286

My research clearly led me to believe that all these phenomena are not explainable with a mysterious *information field* when this information field is seen as functioning independently from the cosmic energy. Such a construct doesn't make sense to me, while we know that the cosmic energy, however we may call it, and that I have come to call *e-force*, as a matter of terminology unification, has more than one characteristic. But it must not be confounded with the energy concept of modern science, which restricts energy to *kinetic energy*. Many conventional researchers namely jump, when they hear the word energy to the assumption the question was about kinetic energy. But this is absolutely not the case. The cosmic energy is not a kinetic energy, not in its pure steady state, as it were, but only after impacting upon matter, in which case, some of the impact can be measured as kinetic force. But the energy as such is a primordial energy field that is informational, and thus there is no need to explain this with quantum considerations. Research on what today is called the *zero-point field*, as it was brilliantly summarized in Lynne McTaggart's book *The Field*, or what Ervin

Published by Sirius-C Media Galaxy LLC, 2011

Laszlo has called the *A-field* in his book *Science and the Akashic Field (2004)* is what I am talking about here. This field has essentially found to be a vacuum, and thus suspiciously reminds of the age-old term *ether*.

Thus, to summarize the book at this point, it can be said that the author has done brilliant research on the verifiability of psychic phenomena without however venturing into the scientific theories or hypotheses at the basis of those phenomena; he has namely left out to clarify what is in last resort the nature of the information field, or what it is that makes that information is transmitted loslessly, over large distances, or in a way that is either to be considered synchronistic or that could be said to proceed at a speed several times the speed of the light. It is my guess that the author intentionally left all this out, to reach more credibility among those who only believe in science when it's measurable, bit by bit and inch by inch. And on the other hand, the book is large enough already, it is true, and I could imagine that those considerations could necessitate a second book on the matter, that the author might bring out in the future. And in fact, it's the meticulous approach of the author that gained him so many merits and convinced so many Cartesian minds of the truth of psychic phenomena, and the author achieved this by a step-by-step procedure in which all parameters are one by one taken up in lab research, and tested beyond reasonable doubt. Such research is very tiresome, and in the face of so much societal resistance, it requires an almost superhuman amount of persistence and faith.

In order to keep this review in reasonable boundaries, I will discuss in more detail here only Chapter 7, entitled *Perception Through Time*, which is an intriguing research topic that was covered also very well in Michael Talbot's study *The Holographic Universe*.

The author explains that the accumulation of conceptual frameworks for explaining how perception and time hang together does not help much, and brings rather more confusion than more clarity. Concepts like *retrocognition*, *real-time clairvoyance* or *precognition*, the author says, rather blur the usual concepts of perception and time. In fact, it is true that they explain nothing. In my view all depends on how we define time, and without doing this, and considering what Einstein said about it, on one hand, and what post-Einsteinian quantum physics says, on the other, we cannot make any valid assumptions about precognition, prophecy or past-cognition.

The author does not advance any theory here, as for example *Michael Talbot* does by explaining all these phenomena with the holographic nature of the universe. When all is one single hologram, it's very well conceivable that the timeline of events is a mere projection system that actually is a crutch for our imagination, while in reality there is no such timeline at all in life, as all events occur simultaneously. But while this concept is very elegant, Radin did not validate or speculate into any of those larger frameworks, and this is again strategically a smart way of doing. For it's easier to bring a theory through that is based on hard facts and verifiable evidence, and later expand the theoretical framework of it; when you do it simultaneously, you put water in your wine, because you get into a generalization level that is beyond what most people can grasp, without having directly verifiable proof to back it up.

Dean Radin does not venture into explanations, here neither, and instead amply discusses research based upon independent replications that he calls *the proof of the pudding* in science. By the way, the book is amply illustrated with graphs and charts.

Published by Sirius-C Media Galaxy LLC, 2011

Dean Radin

Entangled Minds

Extrasensory Experiences in a Quantum Reality
New York: Paraview Pocket Books, 2006

To be frank and straight right at the start, I find this book much more interesting than *The Conscious Universe*, but this has to do with the simple fact that I am researching on these matter since two decades already; for me, the basic proof is since long established.

However, I am well aware that such is not the case with the common lay public, and thus I would recommend *The Conscious Universe* to those who are skeptical, or who are so bare of knowledge of psychic or paranormal events or phenomena that they need to begin with Adam and Eve. Besides, it must be seen that *The Conscious Universe* was published in 1997 and the present book in 2006, that is nine years later. And we all know how much the world has changed in these nine years, because of the present acceleration of events we are experiencing when we get closer to the end of the Mayan calendar, in 2012, as major cosmic and earth cycles are ending, and others restarting at that date.

I shall first make some general remarks about the book, and then focus on Chapter 2 entitled *Naked Psi*, which deals for the most part with the highly intriguing premonitions of the September 11, 2001 events.

Let us first ask, what is entanglement? I find the idea intriguing, not by itself, but because it's not explained as a phenomenon of the cosmic energy, but as a somewhat mysterious condition that has pretendedly nothing to do with energy. Frankly, this reasoning doesn't want to enter my mind. The author writes:

Dean Radin

For centuries, scientists assumed that everything can be explained by mechanisms analogous to clockworks. Then, to everyone's surprise, over the course of the twentieth century we learned that this commonsense assumption is wrong. When the fabric of reality is examined very closely, nothing resembling clockworks / can be found. Instead, reality is woven from strange 'holistic' threads that aren't located precisely in space or time. Tug on a dangling loose end from this fabric of reality, and the whole cloth twitches, instantly, throughout all space and time./2-3

Perhaps, as Dean Radin humbly suggests there is no ready-made answer to this problem, or no answer yet, but he's very optimistic that over the coming years we'll get to see the light on that matter. And I will surely follow through his newer publications. I clearly expect, to be true that the cosmic life energy will be recognized but through the backdoor, and under a new name, and this simply because legions of scientists are afraid they have to eventually give right to those they formerly called *quacks and charlatans*, those who virtually sacrificed their lives and their reputations for the *true science*, people like Paracelus, Mesmer, Reichenbach and Reich; they are afraid that their myopic if not outright fascist attitude do deny what they don't see will be discovered as what it is, or was, and modern science be disqualified and looked upon as a mere trash container.

We all know upon further scrutiny that the Michelson-Morley experiment was a fake and that Einstein was afraid of some heavy impact on his relativity theory in case the ether really existed. And we all know that it exists but to save face, modern science will bring it out under a new header. And that will be it, and we will have finally recognized what is in fact nothing new, but an age-old knowledge that every native tribe on any mountain in the

Published by Sirius-C Media Galaxy LLC, 2011

world possesses since times immemorial. And modern science will have a new occasion to celebrate its great achievements!

Now let me get at the core of this review, the precognitive messages and presentments in the foreground of 9/11, the prophecies as it were not by famous seers but by ordinary people, that were collected in their thousands, as the author reports, by the Rhine Research Center.

The first case he reports was a couple returning from New York to their home town; the man had tried to sleep in the plane, and had a horrible nightmarish vision to be buried alive in tons of cement that were closing hermetically about him, virtually crushing his bones one by one in this prison of stone that was converging about him. When they returned home, exhausted after the long trip and three thousand miles away from their friends in New York, and just went into deep slumber, in New York the two towers of the World Trade Center went down to ashes in an unprecedented catastrophe that was mediatized in its every detail.

In the second documented case, a couple had passed the Pentagon on a highway and the woman, in a sudden vision, had seen the Pentagon burning and huge piles of dark smoke rising from it, while her husband had wondered about her screams. In a few seconds the vision had vanished away.

This had been several weeks before the 11th of September, 2001. The author explains that it is because of the psychological fact of memory repression and a blinding out of perception that so many people do actually not get clear visions; the author actually seems to be sure and convinced that we do receive clear premonitions and visions in front of catastrophic events that cost many human lives, but that our brain safeguards our mental health by suppressing as much as possible of the disturbing impressions and all the anxiety that is of course connected to it.

I'll finalize my review here for the simple reason that it is highly difficult to explain and paraphrase empirical and data-driven research papers and books. It's not only difficult, it also comes often over as amateurish and somehow naive, because truly, if you are really interested in these phenomena it's not enough to rely on what I am saying here, but you have to get the actual book and go through all the data yourself – and I promise you this is not a book to read at bedtime, because if you do it seriously, you have to look up also the notes; in fact, the background research is so gigantic in the meantime that it fills entire libraries. So please believe me that's humility why I stop here instead of wishy-washing over the details, for I can't produce the graphs here and meticulous data for that would be infringing copyright, and besides wouldn't be really useful for it would be too selective and out of context most of the time.

Allow me one remark at the end: it was much more difficult in the past, even a decade ago or so, to write a book on any phenomenon that was covered by parapsychology simply because verifiable data was not easy, if not impossible, to get. And now you have this book and others of this kind where you can quote with the greatest ease, and that is why I think this book belongs in a scientific library, rather than on a bedside table. And there's another consequence of this kind of research. It shall be less comfortable for thousands of rose-colored scam artists to just pretend all and everything under the header of *esoterism*, as we are getting a clearer picture of what is feasible in terms of high human sensitivity, and what is just a matter of belief.

Published by Sirius-C Media Galaxy LLC, 2011

WILHELM REICH

Books Reviewed

Children of the Future (1984)

Cosmic Orgone Engineering (CORE) (1984)

The Leukemia Problem Approach (1951)

The Orgone Energy Accumulator (1951)

The Schizophrenic Split (1945)

I read Wilhelm Reich (1897-1957) early in life, and interestingly enough, before I read Freud. It was when I just had turned eighteen and enrolled in my first semester law school in Germany. Needless to add that I read both Reich and Freud in their German originals, which is not a minor advantage. The German language expresses the *Gestalt* in a much more clear and powerful way than English, and Reich is an author who uses Gestalt thinking and Gestalt-like descriptions all over the place, and that makes him so unique.

Webster's Dictionary

ge·stalt \ge-"stalt, -"shtalt, -"stolt, -"shtolt\ noun pl ge·stalt·en \-"stal-ten, -"shtal-, -"stol-, -"shtol-\ or gestalts [G, lit., shape, form] (1922): a structure, configuration, or pattern of physical, biological, or psychological phenomena so integrated as to constitute a functional unit with properties not derivable by summation of its parts.

Reich was the first researcher in the West who was able to cognitively grasp the true nature of our emotions, as I have pointed this out in more detail in my *Idiot Guide to Emotions* as well as in *The Science of Emonics* and *Emotional Flow*. Significantly so, Reich explains the nature of human emotions as an intelligent function of the bioplasma itself:

Wilhelm Reich

I stress the rationality of the primary emotions of all living. The mechanists of depth psychology have namely spread the view that all emotions were but drives and therefore irrational. However, emotions are specific functions of the protoplasm. Emotions and the natural movement of the bioplasm are functionally identical phenomena.

Regarding the *orgone energy* of which Wilhelm Reich by his present-time hagiographers is always quoted as the unique 'discoverer', the truth is that Reich himself, in his book *Ether, God and Devil,* clearly stated that this allegation was wrong and did not originate from himself or any of his writings.

Wilhelm Reich

It is not correct that it was me who for the first time sighted the orgone energy and thereby discovered the functional law that unifies living and inanimate nature. It was over a course of two thousand years of human history that time and again humans were confronted with manifestations of the orgone energy or they developed systems of thought that were reasoning on the lines of the cosmic life energy. That till now these insights were not officially recognized has its cause in the fact that all progress in this direction was annihilated by those who created religious thought taboos. The forces of destruction always operated either through mechanistic and pseudo-scientific reasoning-away of these facts, or through mystical contempt, if they not proceeded to outright physical destruction.

Source: Wilhelm Reich, *Äther, Gott und Teufel,* Frankfurt/M: Nexus, 1983, pp. 80-81. Translation mine. This book was originally published in its English version: *Ether, God and Devil,* ©1949, 1972 by Mary Boyd Higgins as Trustee of the Wilhelm Reich Infant Trust Fund and published by Farrar, Straus & Giroux, New York. The first edition of *Ether, God and Devil* was published in the English language as Volume 2 of *The Annuals of the Orgone Institute* in 1949.

We know today that scientific genius often manifests by a skill to observe nature in a holistic and systemic way. This is a quality not often to be found with Western scientists, as it was not very much promoted by the former Cartesian science paradigm that was the reigning one in the West until recently, and to a lesser extent in Asia. In fact, manuals of Chinese Medicine or Tibetan Medicine as well as Zen and Taoist writings abound of Gestalt-enriched descriptions of nature.[13] Suffices to read Dr. Reich describing a patient walking in his psychiatric clinic to see how much he could see, in the best sense of the word, of the patient's unique pathology, without yet having done his clinical diagnosis of the patient's psychic health condition. This truly is genius, as we know today, in 2007, at a time where even paranormal healing, and intuitive healing, by people like Caroline Myss, Dora van Gelder-Kunz or Barbara Ann Brennan are more or less accepted.[14] But at Reich's time, this was very different, which is perhaps why he was an outcast almost all through his later years, and died in jail.

If you know these authors, you will understand me when I say that having read Reich before reading Freud has given me an opportunity to read Freud *with critical eyes*. And this has built in me, very truly so, a kind of immunity toward the myth of the *Oedipus Complex* that is unfortunately missing today in most people involved in health care or coaching regarding this greatest myth Freud ever came up with. And I would probably not have built my concept of *Oedipal Culture* if I had not looked through the veil so early in my scientific life.[15]

---

[13] See, for example, Pierre F. Walter, *Alternative Medicine and Wellness Techniques, Scholarly Article (2011)*.

[14] See, for example, Russell E. DiCarlo (Ed.), *Towards A New World View: Conversations at the Leading Edge (1996)*.

[15] Look up the notions of 'Oedipus Complex' and 'Oedipal Culture' in *Walter's Encyclopedia, Academic Edition (2010)* or in *Walter's Encyclopedia, Illustrated Edition, Vol. 1 (Terms)*.

But I owe Reich much more than that. And many mental health profession-als, and coaches today owe him much, without however always admitting it, which is what I find very sad, because silently plagiarizing revolutionary insights from another under a new name or header is not clean scientific behavior.

On the other hand, I would like to state here very clearly that Dr. Wilhelm Reich, against much hearsay, has not invented or discovered the cosmic life energy. Who did, first in Western history, at least as it's documented today, was Paracelsus (1493-1541), and after him Franz Anton Mesmer (1734-1815) and Baron Ludwig Karl Freiherr von Reichenbach (1788-1869), and only thereafter came Reich, at about the same time as Harold Saxton Burr (1889-1973) who did quite a parallel research on what he called the *L-Field*, as Reich with his orgone research, and Georges Lakhovsky (1869-1942) who equally did a parallel research on the cosmic energy and who called it uni-version. And regarding the earliest attempts to healing cancer, there is also a parallel development to note. Not only Reich is famed today to have come up with a unique way to reduce cancerous growth by using bioenergy, but also Lakhovsky published in 1929, in Paris, his book *L'étiologie du cancer* where he shows how he neutralized the bacterial radiations in plants using frequencies, thereby very early establishing what we today call vibrational medicine.

I know that many 'Reichians' do not like to hear this truth and even tend to get very angry at anybody who says that or writes it, but their sectarian zeal is not really bringing the rehabilitation of Wilhelm Reich forward. In the con-trary, their limited and somehow anachronistic worldview sets them apart today as a science clique that is hardly taken serious anywhere in the world. And their fixation on the orgone as the only valid set of research parameters to inquire in the functionality of the cosmic life energy is simply ridiculous. Where is there sense for diversity? Why are they so hard on reproaching current science to cast out the energy formula, while they themselves remain stuck on a single term, called *the orgone*, and don't see that what other scientists have named differently is basically the same soup?

As a matter of fact, current science, as by 2008, has accepted the cosmic life energy as a valid ingredient in Western science, but under a quite different name. It's called alternatively either the *zero-point field*, for example by Ervin Laszlo or Lynne McTaggart, or the *quantum vacuum*, in the words of Dr. William Tiller. Here are some of the leading-edge publications that show that basically the existence of the ether and of the cosmic life energy are today accepted and integrated in the new holistic science paradigm, as it is the final outcome of the strongly shattering impact of quantum physics on the formerly Cartesian science paradigm:

McTaggart, Lynne
The Field
The Quest for the Secret Force of the Universe
New York: Harper & Collins, 2002

Laszlo, Ervin
Science and the Akashic Field
An Integral Theory of Everything
Rochester: Inner Traditions, 2004

Tiller, William A.
Conscious Acts of Creation
The Emergence of a New Physics
Associated Producers, 2004 (DVD)

Goswami, Amit
The Self-Aware Universe
How Consciousness Creates the Material World
New York: Tarcher/Putnam, 1995

Published by Sirius-C Media Galaxy LLC, 2011

Talbot, Michael
The Holographic Universe
New York: HarperCollins, 1992

Karagulla, Shafica
The Chakras
Correlations between medical science and clairvoyant observation
With Dora van Gelder Kunz
Wheaton: Quest Books, 1989

Sheldrake, Rupert
A New Science of Life
The Hypothesis of Morphic Resonance
Rochester: Park Street Press, 1995

Todaro-Franceschi, Vidette
The Enigma of Energy
Where Science and Religion Converge
New York: Crossroad, 1999

But of course, I am not reductionist nor am I pro or against Reich, as strangely enough so many people, as Reich seems to divide the minds around his charismatic figure. I would like to emphasize that all of the polemics pro or con Reich should not keep anyone from reading Reich because to read the books of a genius scientist never is superseded by newer scientific developments, nor is it in any way distorted by the personality of the scientist. Reich's books are very well written, very honestly researched, very sound and logically meticulous, and they abound of clinical examples. They are not at all the writings of a fanatic, as some of his detractors would like to let them appear in the great public, but of a sane mind. Reading Reich is highly educational in many respects! It has been for me, and I am thankful for that. To become fanatic about Reich and his life, and put up hagiographies about him is not very scientific, to be true! And it's not in the sense of Reich's idea of scientific heritage. Reich – a saint! He would have felt offended, surely, by this kind of productions, and misunderstood.

And what I found in years of biographic research on Reich was clearly that Reich's style and manner to treat people, to deal with relationships, and ultimately to handle or mishandle his emotions was a major causal factor in the contempt he was triggering in others, and in authorities. Ilse Ollendorf-Reich, his second wife, has given some scant evidence of his character in her biography. Ilse Ollendorf-Reich writes:

> **Ilse Ollendorf-Reich**
>
> Aber auf der anderen Seite konnte er ein strenger, ungeduldiger, furchteinflössender Mann sein, besonders bei seinen Assistenten und Mitarbeitern. (…) Er drängte Menschen oft / erbarmungslos und verlor viele gute Mitarbeiter, weil sie mit ihm nicht Schritt halten konnten.
>
> But on the other hand he could be a hard, impatient and frightening man, especially with his assistants and collaborators. (…) He often ruthlessly bullied people and thereby lost many good collaborators, because they couldn't catch up with him.
>
> **Source**: Ilse Ollendorf Reich, *Wilhelm Reich*, Vorwort von A.S. Neill, München, Kindler, 1975, pp. 19-20. (Translation mine).

Reich's greatest discoveries cannot be wiped under the carpet. Reich has found an effective way of treating cancer, long before the French couple Dr. Carl O. Simonton and Dr. Stephanie Matthew-Simonton gained their merits with non-mainstream cancer healing. And his orgone research has shown that DOR-affected emotions can be carefully retransformed into healthy functional emotions by muscular body work, in just the same way as in Feng Shui, *sha* is retransformed into lively *ch'i* by using a remedy. But in my view Reich's greatest merit is to have proven that schizophrenia really can be healed with orgonotic treatment, and this is a medical revolution that searches its equal. In his case report *The Schizophrenic Split (1945/1949/ 1972)*, Wilhelm Reich stated:

Wilhelm Reich

The general deterioration of the organism in later phases of the process is due to chronic shrinking of the vital apparatus, as in the cancer biopathy, though different in origin and function. The shrinking carcinomatous organism is not in conflict with social institutions, due to its resignation. The shrinking schizophrenic organism is full of conflicts with the social pattern to which it reacts with a specific split./36

While most of Reich's scientific heritage is still scarcely represented in modern research, I have chosen for review five of the lesser known books by Wilhelm Reich. These books, partly because they either appeared rather late in mainstream publishing media or are still only available as XEROX copies from the Wilhelm Reich Museum, are among the best that Reich has left us from his huge scientific legacy. Some of the insights Reich developed over the course of his life as a physician, psychoanalyst and bioenergy researcher, mainly in his book on the prevention of sexual pathology, *Children of the Future*, are so important that their continued neglect in scientific and political policy-making requires a particularly heavy tribute, the safety and well-being of children in all Western industrialized nations. The rampant sexual violence against children, especially in countries like the United States, while being recognized as a major problem, is hardly ever fought with today's ineffective and inappropriate means that the current political and legislative power structures allow and endorse.

Not only Reich's books, but also many of the newer sociopolitical studies in sex politics and the roots of violence, such as, for example, the research of

James W. Prescott, Ashley Montagu, Frederick Leboyer, Michel Odent or Alice Miller deliver evidence for the fact that endlessly tightening criminal laws does not produce any positive result in reducing sexual violence targeted at children, while it makes long-term for a climate of lynch justice, insecurity, persecution and violence not only in the United States, but, as the USA exports their paradigms worldwide, also in many other industrialized nations.

The solutions to these highly complex problems cannot come as a fortunate strike of heaven, but will, if ever, be the result of careful analysis and cross-disciplinary synthesis of research results, and this beyond national borders, and through an effort of international or supranational cooperation. Elements of this metarational and synthetic effort shall be:

- A total reform of our educational system;

- A reform or complete abolishment of so-called 'morality';

- The implementation of a permissive education that allows children to live their loves;

- A total abolishment of so-called 'sex laws', i.e. the laws of 'age of consent';

- A total abolishment of all 'sex laws', giving the citizen the freedom to responsibly live his or her emotional attractions and sexuality, without the nation states and their police or security forces to interfere in any way;

- The setup of consulting agencies or authorities that are legally empowered to advise and consult the citizen about responsible, healthy, constructive and non-harmful forms of emotional and sexual bonding.

Published by Sirius-C Media Galaxy LLC, 2011

Wilhelm Reich

Children of the Future

On the Prevention of Sexual Pathology
New York: Farrar, Straus & Giroux, 1984
Preface © 1983 by William Steig

Children of the Future was one of the books that Reich wrote late in his life, and it therefore condenses many of his fundamental insights in the natural unbent and unarmored human nature. What mainstream Western medicine hardly ever did, Reich provided in his approach to health from the start of his career: concise ideas and advice for the prevention of pathology, and a solid base knowledge of what *health actually is*, both physical and psychic health, and per analogy, what unperverted human sexuality looks like. And from this base paradigm of health, Reich could formulate a sound catalog of ideas about how to setup a future educational system that would insure that children's emotions and sexuality are not thwarted into denial, violence, perversion and sadism. With astounding honesty and clarity of mind, Reich states the status quo at the beginning of his book:

Dr. Wilhelm Reich

We are no more than transmission belts from an evil past to an eventually better future. We shall not be the ones to build this future. We have no right to tell our children how to build their future, since we have proved unfit to build our own present. (...) We cannot possibly preach cultural adaptation for our children when this very same culture has been disintegrating under our feet for more than thirty-five years. Should our children adapt to this age of war, mass killing, tyranny, and moral deterioration?/6

In my entire life I haven't encountered such an honest statement about the state of the world and what we can do about it – facing the unfulfilled needs of our children. When Reich wrote this, in the 1950s, the world was by far not as catastrophic, and ecologically devastated as it is by now, in 2008, and the whole of the Western world was by far not as fascist and pleasure-denying toward children than this is the case today. Hence, the actuality of Reich's book is even greater today. Our pathologies, and our sexual pathologies have not seen an end in these fifty years of ongoing madness, and our legislators have not done one single move in the right direction to end the sexual misery of the youth – in the contrary, all has only worsened. And as to the following of Reich's statements, I really wonder where we are, characterologically, today:

> **Dr. Wilhelm Reich**
>
> The structural distortions in the character of the parents, physicians, and educators are transmitted automatically to the next generation./9

Reich's view of preventing sexual pathologies in child rearing is essentially one that fosters *non-interference* in children's emotional and sexual life. It is a view that admittedly today, and fortunately so, is shared by many progressive child psychologists, psychoanalysts, and social workers, and yet it is not for this reason a mainstream view, and it has not for this reason gained access to any of our arrogant and ignorant governments that go on to mold children after certain standards, certain values, certain ideologies, certain religions, certain racial or otherwise partial views, certain fashions and certain traditions. The view that children should be left to grow by themselves is today as marginal as it was in the 1950s. And therefore, I have nothing to add to what Reich wrote:

Dr. Wilhelm Reich

The human species has for millennia been split into numerous groups according to nationality, race, religion, state, etc. Each group has directed its educational measures toward the adjustments of every new generation to the specific national, religious, or racial ideals and institutions. A dictator, if asked what he thinks a healthy child should be like, would doubtless say he should be a good defender of the honor of the fatherland. A Catholic would claim that a healthy or so-called normal child is one who obeys the customs of Catholicism; killing the ‚sinful' craving ‚of the flesh' appears to be the main criterion. The member of Western civilization would define the healthy child as the ideal bearer of Western culture, and the representative of Eastern culture would, by the same token, define health in the child as the ability to be obedient, stoical, unemotional, and fit to carry on the old traditions of the Eastern patriarchate. The official view in dictatorial Russia is that the child ‚should be like Stalin'. We, on the other hand, do not want our children to be like Stalin, or like anybody else for that matter. We want them to be themselves./14

Reich doesn't stay with generalities, but gives a detail account of what it is that distorts the intrinsic nature of the child within our educational system, and that has not changed since Reich's tragic death in jail, in 1957. Reich explains that there are essentially two fundamental wrong-doings in our child-rearing practices that contribute to thwart and distort the originally sane psychosomatic setup of the child. Here they are in his own words:

Dr. Wilhelm Reich

• The natural bioenergetic principle in the newborn baby is systematically smothered and ruined by the armored parent and educator, who in turn are supported in their ignorance by powerful social institutions which thrive on the armoring of the human animal.

• A simple but tenacious misinterpretation of nature governs all education and cultural philosophy. It is the idea that nature and culture are incompatible. In accordance with this cultural ideology, psychoanalysts have failed to distinguish between primary natural and secondary perverse, cruel drives, and they are continuously killing nature in the newborn while they try to extinguish the brutish little animal. They are completely ignorant of the fact that it is exactly this killing of the natural principle which creates the secondary perverse and cruel nature, human nature so called, and that these artificial cultural creations / in turn make compulsive moralism and brutal laws necessary./17-18

Reich vehemently contradicted the mechanistic approach of most Western physicians and pediatricians, which was one of the reasons why he did not gain social acceptance throughout his life. His greatest enemies were not governments or secret services, but his own colleagues. Contrary to common belief, Reich maintained an excellent relationship with the federal government after his emigration to the United States, and as I shall report it in my review of *Cosmic Orgone Engineering*, Reich has not failed to alert and inform the US Air Force of his astounding discoveries regarding desertification and DOR accumulation in certain regions of the United States, thus doing all he could for public health, as a good and responsible scientist-citizen. That Reich was imprisoned was because of an injunction of the FDA against the commercial distribution of orgone accumulators, and the FDA, as everybody knows, is primarily driven and run by the medical establishment. Reich has unveiled many of the erroneous beliefs the medical establishment was suffering from still in the 1950s, while in the meantime much has changed in this respect, not only in the United States. Reich writes:

Dr. Wilhelm Reich

If no severe damage has already been inflicted on it in the womb, the newborn infant brings with it all the richness of natural plasticity and development. This infant is not, as so many erroneously believe, an empty sack or a chemical machine into which everybody and anybody can pour his or her special ideas of what a human being ought to be. It brings with it an enormously productive and adaptive energy system which, out of its own resources, will make contact with its environment and begin to shape that environment according to its needs./20

It is obvious for everyone who is informed about the ways of life and the child-rearing practices of native cultures that basically mother nature has provided for all, and that we therefore do not need expensive machinery for giving birth, and for providing medical service.

Wilhelm Reich

Cosmic Orgone Engineering (CORE)

Part 1 : Space Ships, DOR and Drought
Publications of the Orgone Institute
Vol. VI, Nos. 1-4, July 1984
Orgone Institute, Rangeley, Maine, 1984
XEROX Copy distributed by the Wilhelm Reich Museum
http://www.wilhelmreichmuseum.org/books.html

Cosmic Orgone Engineering (CORE) is a sampler volume that contains four issues of the Publications of the Orgone Institute, from July 1954. The subject matter of this 140 pages DIN A 4 book is what Reich called OROP Desert. As it is not self-evident what Reich meant by this expression, here is what he explains about the term in the Introduction:

Dr. Wilhelm Reich

The story of OROP DESERT is long and complicated. Let us begin with the word OROP. This word was coined to designate all operations on the part of human beings regarding DESERT. Desert here includes drought, atmospheric conditions which lead to drought and desert, and the technical means, based on the scientific understanding of nature, which could, possibly, do something beneficial about the prevention of drought and desert development. 'OR' is included because the scientific data underlying our technological data on desert development have been worked out in the realm, method, research and technical development of thought, which differs from other systems of thinking in that it is neither mechanistic nor mystical, but functional, energetic. Thus, 'OROP' comes to designate engineering operations involving the COSMIC ENERGY FUNCTION./v

To begin with, Reich's idea of desert follows an interesting line of thinking that observes a coincidental behavior between atmospheric desertification and emotional shrinking, the result of millennia of emotional and sexual repression under patriarchy - that Reich metaphorically calls another form of desertification. I find this idea unique and have encountered it nowhere else. And from his reasoning, two lines of research may ensue, or topological approaches evolve. One would ask if outward desertification could possibly lead to, or contribute to bringing about emotional desertification? The other approach would ask if it was possible that emotional desertification would have an impact on the environment in the form of an energetic interaction of our bioplasma with the atmospheric orgone that leads to outward desertification?

Both scientific theories, that are left open by Reich as he came up with the desert research quite late in life, are debatable, but have to my knowledge never been followed up to by any scientist. Those scientists who erected the theory that patriarchy resulted from atmospheric desertification in North Africa should have asked the pertinent question if not that atmospheric desertification was in turn a result of the *emotional desert* of the people under patriarchy, within that region? If the latter could be scientifically proven, this would namely put their theory upside down.

Was patriarchy a result of desert? In my view, desert was a result of patriarchy - and not the other way around. And from the present and other sources in the larger context of Reich's research on desert, while much of this material are still undisclosed, I believe that the second line of research would be more fruitful than the first, that is to prove that desert is the result of *emotional shrinking within the humans* that populate the regions that shrinks into desert. Reich's most daring scientific hypothesis is that UFOs contribute to the desertification of the globe because of the *Melanor* traces Reich found they leave wherever they appear. This is to my knowledge a hypothesis that no other researcher has ever come up with, while in many UFO sightings indeed a sickening kind of atmosphere around the UFO was reported. Reich writes:

Dr. Wilhelm Reich

It was clear from the very beginning of the DOR emergency that we were dealing with a functional process which somehow converted cosmic energy directly into matter-like substance, and also the opposite way. These functions were assumed to be operating below the realm of mechanical, electrical and chemical functions, as pre-atomic, sub-chemical, primordial functions of the universe. Therefore, in July 1953, a new branch of chemistry was inaugurated at Orgonon. Its objective was to clear up the pre-atomic chemistry and bio-chemistry involved in ORANUR. The result of this work was the discovery of a white, substance-like, but pre-chemical matter which was termed ‚ORENE'. It contained no less than the principle of growth per se in the form of a white (yellow in acid) powdery substance./10

Wilhelm Reich

The Leukemia Problem Approach

© 1951 Orgone Institute Press
© 1979 Mary Boyd Higgins as Trustee of the Wilhelm Reich Infant Trust
XEROX Copy distributed by the Wilhelm Reich Museum
http://www.wilhelmreichmuseum.org/books.html

THE LEUKEMIA PROBLEM: APPROACH

WILHELM REICH

The Leukemia Problem is a tiny four-pages publication distributed by the Wilhelm Reich Museum that deals with a specific kind of cancer, so-called blood cancer or Leukemia. Based on his general insights about an alternative form of cancer therapy using orgone energy accumulated in the *Orgone Energy Accumulator* and similar devices, Wilhelm Reich was writing this small paper in order to explain that Leukemia, too, could be tackled using the orgonotic approach to healing. Reich explains:

Dr. Wilhelm Reich

In Cancer research it had been established beyond any shadow of a doubt that the formation of protozoal Cancer cells in tissues was itself a secondary process, a reaction, as it were, to a deeper and more basic disease process. At the core of this primary disease process, orgonomic Cancer research found a general weakness of bio-energy, a low charge and continuous loss of bio-energy in the tissues and the RBC, or most outspoken in certain localizations, usually where muscular armoring and blocking of movement of bio-energy had choked off normal tissue functioning and blood circulation. Before Cancer cells developed, which usually happens late in the total Cancer process, there was at work for many years, more often for decades, a slow dying process, a decay and putrefaction of tissues and blood due to characterological, bio-energetic withdrawal and resignation and to putrefaction. The so-called T-bacillus, which gained such overall importance in the understanding of the Cancer process as the years passed by, can be obtained from any kind of tissue or protein through degeneration and putrefaction. It is now this „T-REACTION" which provokes the development of PA bions from which in turn Cancer cells arise. Thus, in the last analysis, the Cancer cell itself is a RESULT rather than a cause of the Cancer process, though, it is also true that the Cancer cells through rapid growth, infiltration of organs, and consecutive decay enhance the dying process./79

Published by Sirius-C Media Galaxy LLC, 2011

Wilhelm Reich

The Orgone Energy Accumulator

Its Scientific and Medical Use
© 1951 Orgone Institute Press
XEROX Copy distributed by the Wilhelm Reich Museum
http://www.wilhelmreichmuseum.org/books.html

The Orgone Energy Accumulator is a booklet distributed by the Wilhelm Reich Museum that over 56 pages explains the functioning and use of Reich's main orgone energy accumulating device. While other devices are mentioned, such as for example an orgone energy blanket and an orgone energy shooter, an orgone energy funnel for local application and treatment, the book provides scientific data, illustrations (photos) and explanations mainly on the accumulator only. The booklet comes with a handy and valuable little bibliography of not only Reich's but also other medical doctors' and researchers scientific evaluations of the accumulator.

Please note that what is often sold in popular magazines and on the Internet as miracle devices and that might look like Reich's orgone energy accumulator, may have nothing to do with it, or may even be based on fraudulent marketing. In order to be sure if any such device is really an *Orgone Energy Accumulator* that is built from Reich's own plans and ideas, you have to investigate the matter.

Let me provide only two quotes from this booklet. The first quote I am going to provide and that was probably written by the publisher is quite important as it shows that Dr. Wilhelm Reich was not unaware of the bioenergy research that has been accomplished before his own discovery of the orgone. Hence, those who write in their publications that Reich had erected himself to be the first and only discoverer of the cosmic life energy are wrong with their allegations:

Dr. Wilhelm Reich

The existence of a specific kind of energy which directs and maintains LIFE has been theoretically assumed for a long time by many natural philosophers and scientists. Wilhelm Reich succeeded in / connecting known facts and in finding new facts which he coordinated with the known facts in such a manner that the existence of a specific life energy became concretely demonstrable, usable, and measurable. ORGONE ENERGY is the name given by Wilhelm Reich to this specific life energy WITHIN the living organism. The existence of orgone energy OUTSIDE the living organism, in the atmosphere, was also demonstrated. The INNER (life-) energy derives from the OUTER orgone energy in the atmosphere. This is quite logical, since nothing can exist within the living organism that did not previously exist in the environment of the living organism./12-13

The second quote that I would like to provide was apparently written by Wilhelm Reich himself. He discusses here some of the more notorious ways how his research was seen from the side of the science establishment, and how little it had been validated by what he calls 'the shapers of public opinion.'

Dr. Wilhelm Reich

Each single new fact which revealed itself to the orgonomic researcher seemed to contradict most cherished scientific beliefs of long standing, as, for instance, the air germ theory in bacteriology; the theory of 'static electricity' in electrophysics; the theory of 'heat waves' in meteorology; the theory of 'cosmic radiation' allegedly coming down to us from faraway 'empty space'; the interpretation / of the many varied and lawful functions of the spontaneous electroscopic discharges as 'mere' results of an incomprehensible 'natural leak'; the Second Law of Thermodynamics with its entropy; the theory of atoms and electronic particles as the primal constituents of the universe; the utter awe which governed any and every thought of a concrete, measurable, and actionable life energy; the clear-cut microscopic observations which disclosed the natural organization of single-cell organisms from decaying tissue and even from free orgone energy in frozen, highly-charged water and which so sharply refuted the basic trends in present-day biology; and many, many similar facts, each single one of cosmic dimensions. (...) However, the basic findings had been published since 1938 in a / continuous stream of publications which were available in many libraries all over the world, but which were little appreciated by the shapers of public opinion./10-12

Wilhelm Reich

The Schizophrenic Split

Its Scientific and Medical Use
© 1945, 1949, 1972 by Mary Boyd Higgins as Trustee of the Wilhelm Reich Infant Trust
XEROX Copy distributed by the Wilhelm Reich Museum
http://www.wilhelmreichmuseum.org/books.html

The Schizophrenic Split is one of Wilhelm Reich's most important publications, while it is hardly known. This lengthy case report of treating a highly delusional schizophrenia patient orgonotically over several months was closed successfully - and this alone is a revolution in medical science. According to this report, Reich really was able to heal this patient completely from acute schizophrenia, despite the fact that the treatment was at times turbulent and that there were many drawbacks, including suicide attempts of the patient. I shall provide some comments on this extraordinary book and some quotes.

I have mentioned in my review of Jung's paper *On the Psychogenesis of Schizophrenia* that I would like to rectify some of Jung's wrong or superficial assumptions about schizophrenia, especially when compared to Wilhelm Reich's extensive research and successful orgonotic treatment of an illness that is caused by insufficient bioenergetic and emotional flow.

To begin with, schizophrenia is by no means a mental illness. It does in no way affect the rational mind and intelligence of the subject. By contrast, emotionally the subject is stuck. There are areas in the mindbody of the schizophrenic that lack bioenergetic charge and that thus are in a state of reduced vitality. But contrary to the cancer etiology, Reich makes it very clear in the present case report that schizophrenia is not an etiology where tissues are in a state of biogenic retardation and shrinking, but where the person has lost the basic touch with her body sensations, thus projecting a distorted meaning in those sensations. The lack of vitality comes about through a distortion of perception that affects the body negatively through a general negativity of the mind regarding the basic life functions. The schizophrenic, for example, speaks of *The Forces* when they perceive their bioenergetic streaming in the body. Because of early negative conditioning, they have learnt to regard their natural hot and melting sexual emotions as alien to their organism, or mindbody; that is why their natural primary power is felt as an outside force that commands them to do certain things, indulge in certain obsessions or even murder people *on command*.

Contrary to Jung's view, that well is novel in that is posits the psychogenic etiology of schizophrenia versus the former myth that schizophrenia was the result of genetic or physical brain damage, Reich, and here he is really on a singular path of exploration, asserts that schizophrenia is caused by an acute lack of bioenergetic flow. Reich states:

Dr. Wilhelm Reich

Thus, a schizophrenic will fall into a state of disorientation when his self-perception is overwhelmed by strong sensations or orgonotic plasma streamings; the healthy genital character will feel well, happy, and highly coordinated under the impact of orgonotic streaming. (...) Our approach to schizophrenia is a biophysical, and not a psychological one. We try to comprehend the psychological disturbances on the basis of the plasmatic dysfunctions; and we try to understand the cosmic fantasies of the schizophrenic in terms of the functions of a cosmic orgone energy which governs his organism, although he perceives his body energy in a psychotically distorted manner. (...) The general deterioration of the organism in later phases of the process is due to chronic shrinking of the vital apparatus, as in the cancer biopathy, though different in origin and function. The shrinking carcinomatous organism is not in conflict with social institutions, due

> to its resignation. The shrinking schizophrenic organism is full of conflicts with the social pattern to which it reacts with a specific split./35-36

The whole drama, the fatal truth is that schizophrenics perceive their own bioenergetic streaming as an *alien energy* that they dissociate from their perception. It is this early dissociation of body sensations and meaning, which often was conditioned as early as in childhood, that is the essential characteristics in the etiology of schizophrenia.

I do not know any researcher, from which discipline he or she may be, who has found this essential etiology of schizophrenia and who was also able to heal the illness when it was well established over years and years already. Reich has this merit, while to my knowledge he was never credited with it. This is not so surprising after several decades of slander and media manipulation that he has encountered, even after his death in prison, and also because of the fact that this booklet is not available in ordinary publishing, but only as a XEROX copy from the Wilhelm Reich Museum in Maine.

Published by Sirius-C Media Galaxy LLC, 2011

CARL REICHENBACH

Baron Ludwig Karl Freiherr von Reichenbach (1788-1869), who was a recognized chemist, metallurgist, naturalist and philosopher, and a member of the prestigious *Prussian Academy of Sciences*, known for his discoveries of kerosene, paraffin and phenol, spent the last part of his life observing the vibrational emanations and the energy code in plants. He spoke of *Od* or *Odic force*, a life principle which he said permeates all living things and that before him was discovered by Paracelsus and Mesmer.

Reichenbach was by no means a mystic or mystical thinker, but throughout his life a natural scientist. All his conclusions were based on the controlled observation of natural processes, especially in plants and in humans and the interactions between plants and humans. For example, when observing a plant in a darkened room in the cellar of his castle that he had isolated against telluric vibrations, he observed, after having accustomed his eyes to the complete dark for about two hours, a blue-green shadowy egg-formed substance around the plant. After having been certain about his own accurate perception and repeatability of the experiment, he invited other scientists and lay persons to join him in his observations, and those other persons, who were carefully selected in terms of mental clarity and sanity, corroborated his observation.

On the basis of his observations, Reichenbach set out to heal sick people with the Odic force construing various devices for this purpose. He became very popular as he, as a very rich industrial, went to the poor in order to heal their suffering family members.

Reichenbach's research corroborates a part of the spiritual microcosm of the native Kahunas in Hawaii and the corresponding cosmology of the Cherokee natives in North America who almost exclusively use plant-contained bioenergy in their approach to heal disease.

Published by Sirius-C Media Galaxy LLC, 2011

RUPERT SHELDRAKE

A New Science of Life

The Hypothesis of Morphic Resonance
Rochester: Park Street Press, 1995

LOOK **INSIDE!**™

A New Science of Life, by Rupert Sheldrake is a book has received much criticism, and I suppose not because it's badly written, but because scientists disagree with most of the conclusions of the author. I must make it very clear that my critique is of a different nature; in fact, the reasons of my criticism are for the most part the exact opposite of the critique this research received in the world of established science since its first publication in 1981. While the tenor of this criticism is that Sheldrake went too far, the tenor of the present review of his research is that he did not go far enough, and that he actually hangs and struggles in his own web of blind spots he never seems to overcome, and this for one single major reason. I may be allowed to illustrate my point by putting up the following little question to the author, as a sort of public letter to the author:

Public Letter to Dr. Rupert Sheldrake

Dear Sir,

I know all you say is true except where you say the morphogenetic germ is not a matter of energetic impact, but a cause of its own. May I know why you replace life and life energy by the term vitalistic theory? Is life a theory for you? And besides, where do you think the memory of the pattern that emerges in the form of a morphogenetic resonance is stored if not in the Akashic field, that is in cosmic life energy that constitutes the ether, L-Field, orgone or however you may call it? It is clear that you can't have a valid answer here for as long as you exclude the cosmic life energy from your conclusions, you can never adequately describe nature. You did not stray too far from the science pack, Mr. Sheldrake, but not far enough to see that you are still under the spell of the mechanistic, lifeless, vivisectionist and non-organic scientific worldview of the Cartesian age. For if you'd see that the life energy is the carrier of all memory phenomena in life, be it with water, in the form of hado, as Masaru Emoto showed it and other water research, or in any of the multiple forms in nature, then you would not need at all the construct of a morphogenetic field, or you would call this field what it actually is: an energy field or energy pattern. But you say that it's not energy. It's as if I was saying Mr. Sheldrake is not a living human. So how tell me how it comes that Mr. Sheldrake writes books and does research if he's not animated by the cosmic energy? Oh, it's certainly because he's animated by morphogenetic fields as his personal life battery, right?

And that is why I think this is a dead theory, and not, really not, *A New Science of Life*, as the book is euphemistically entitled.

I ordered this book with much expectations but as always, the reader views on amazon.com were completely misleading and none of them had looked deep enough to see the spook of this book, and the for the most part untenable scientific conclusions of its author. The joke of this book is that there are no conclusions at all. The author leaves his audience with *Four Possible Conclusions* in Chapter 12. That may be a good trick, after all, that is to leave all doors open in case I've done a major mistake here, they won't catch me. I always have a backdoor.

In the subject matter of his research the author is not less evasive, while implicitly and as a matter of terminology used, the non-stupid reader quickly sees behind which monster the author hides, and why.

Published by Sirius-C Media Galaxy LLC, 2011

In the descriptive parts, the books is well-written, as for example in the following passage, where the author gives a short overview over the basic understanding underlying the holistic worldview. He writes:

Rupert Sheldrake

The organismic or holistic philosophy provides a context for what could be a yet more radical revision of the mechanistic theory. This philosophy denies that everything in the universe can be explained from the bottom up, as it were, in terms of the properties of atoms, or indeed of any hypothetical ultimate particles of matter. Rather, it recognizes the existence of hierarchically organized systems which, at each level of complexity, possess properties which cannot be fully understood in terms of the properties exhibited by their parts in isolation from each other; at each level the whole is more than the sum of its parts. These wholes can be thought of as organisms, using this term in a deliberately wide sense to include not only animals and plants, organs, tissues and cells, but also crystals, molecules, atoms and sub-atomic particles. In effect this philosophy proposes a change from the paradigm of the machine to the paradigm of the organism in the biological and in the physical sciences./12

I do not quite understand the hypothesis itself. It appears to be well put in words, but it sounds insignificant. If I stay with my conflicting interpretation of morphogenetic fields being energy patterns, then I can of course agree with the author that they are going to result in measurable physical reactions. Now what the hypothesis says further is that the forms, all forms in nature are determined by these fields, but I do not see why for that reason they should not be energy fields but fields sui generis. I also agree with the author that these fields may be specific in that they may restrict 'energetically possible outcomes', but why then, are they themselves not related in any way to the life energy?

Rupert Sheldrake

The hypothesis put forward in this book is based on the idea that morphogenetic fields do indeed have measurable physical effects. It proposes that specific morphogenetic fields are responsible for the characteristic form and organization of systems at all levels of complexity, not only in the realm of biology, but also in the realms of chemistry and physics. These fields order the systems with which they are associated by affecting events which, from an energetic point of view, appear to be indeterminate or probabilistic; they impose patterned restrictions on the energetically possible outcomes of physical processes./13

From what Sheldrake found, morphogenetic fields are but patterned structures; the contain energy molds, that is how I see it. They are responsible for molding form in nature, thus they contain a code. But Sheldrake is merely descriptive here in that he does nothing for elucidating what this code is in fact. When he says the structures are from similar patterns from the past, he brings in a time element, and accordingly memory but without ever mentioning that what we have here is a memory effect.

Rupert Sheldrake

If morphogenetic fields are responsible for the organization and form of material systems, they must themselves have characteristic structures. So where do these field-structures come from? They answer suggested is that they are derived from the morphogenetic fields associated with previous similar systems: the morphogenetic fields of all past systems become present to any subsequent similar system; the structures of past systems affect subsequent similar systems by a cumulative influence which acts across both space and time./13

According to this hypothesis, systems are organized in a way they are because similar systems were organized that way in the past. For example, the molecules of a complex organic chemical crystallize in a characteristic pattern because the same substance crystallized that way before; a plant takes up the form characteristic of its species because past members of the species took up that form; and an animal acts instinctively in a particular manner because similar animals behaved like that previously./13

A number of testable predictions can be deduced from this hypothesis which differ strikingly from those of the conventional mechanistic theory. A single example will suffice: if an animal, say a rat, learns to carry out a new pattern of behaviour, there will be a tendency for any subsequent similar rat (of the same breed, reared under similar conditions, etc.) to learn more quickly to carry out the same pattern of behavior. The larger the number of rats that learn to perform the task, the easier should it be for any subsequent similar rat to learn it. Thus, for instance, if thousands of rats were trained to perform a new task in a laboratory in London, similar rats should learn to carry out the same task more quickly in laboratories everywhere else. If the speed of learning of rats in another laboratory, say in New York, were to be measured before and after the rats in London were trained, the rats tested on the second occasion should learn more quickly than those tested on the first. The effect should take place in the absence of any known type of physical connection or communication between the two laboratories./14

This hypothesis, called the hypothesis of formative causation, leads to an interpretation of many physical and biological phenomena which is radically different from that of existing theories, and enables a number of well-known problems to be seen in a new light. In the present book, it is sketched out in a preliminary form, some of its implications are discussed, and various ways in which it could be tested are suggested./14

I really do not understand why an intelligent researcher such as Sheldrake needs to get beaten in the ring by the mechanistic morons that populate our Western science establishment. After all, this book is but a defense against mechanism, but Sheldrake thinks he can get out of the ring with a blue eye by making down the truth of energy, cosmic life energy being the (only) memory surface that is thinkable here, calling the truth *vitalism*. He was anyway beaten up and his book was suggested for book burning - which does not astonish me for the least. It was a good test forum to see what spirit reigns especially in Anglo-Saxon scientific rings, a spirit namely which is hardly advanced over the Church's dogmatic views of the 15th century; symptomatically so, the very idea of book burning is one of the most flagrant pieces of evidence for such a spirit.

But why, to repeat it, is Sheldrake advancing a theory that is in last resort untenable because it sits exactly between the chairs? It sits between the chairs of materialism, on one hand, and of a spiritual worldview, on the other. It denies the creator principle, but it also denies materialism, at least implicitly, while Sheldrake carefully tries to avoid any position pro or con. But in my view, his weakness is exactly his lacking outspokenness as to the ultimate position he takes in the scheme of things, and regarding to the explanation of cosmic memory, a point that for example the Hungarian researcher Ervin Laszlo has clearly answered to the affirmative in his book *Science and the Akashic Field (2004)*.

Rupert Sheldrake
The discovery of the structure of DNA, the cracking of the genetic code and the elucidation of the mechanism of protein synthesis seem to be impressive confirmations of the validity of the mechanistic approach./17

I think that Fritjof Capra has done such a good job in summarizing the valid criticism of the mechanistic approach. Sheldrake could have safely relied on that, instead of venturing in an open fight with the stupidest and most stubborn of all possible scientists, that is the materialists.

Published by Sirius-C Media Galaxy LLC, 2011

There's no doubt that the formative structures or molds are not just developmental but epigenetic in the sense that they contain novelty content, and here again, Sheldrake's research is merely descriptive, and the same is true for regeneration. These insights are not even new, they are age-old, but Sheldrake completely misses out to look for the origin of these phenomena, which can only be traced back to the cosmic memory matrix that is inherent in the cosmic life energy.

Rupert Sheldrake

The first problem is precisely that form comes into being. Biological development is epigenetic: new structures appear which cannot be explained in terms of the unfolding or growth of structures which are already present in the egg at the beginning of development. The second problem is that many developing systems are able to regulate; in other words if a part of a developing system is removed (or if an additional part is added), the system continues to develop in such a way that a more or less normal structure is produced./19

The third problem is that of regeneration, whereby organisms are able to replace or restore damaged structures. Plants show an amazing range of regenerative abilities, and so do many of the lower animals; if a flatworm, for example, is cut up into several pieces, each can regenerate into a complete worm. Even many vertebrates possess striking powers of regeneration./20

The hypothesis of formative causation proposes that morphogenetic fields play a causal role in the development and maintenance of the forms of systems at all levels of complexity. In this context, the word form is taken to include not only the shape of the outer surface or boundary of a system, but also its internal structure. This suggested causation of form by morphogenetic fields is called formative causation in order to distinguish it from the energetic type of causation with which physics already deals so thoroughly. For although morphogenetic fields can only bring about their effects in conjunction with energetic processes, they are not in themselves energetic./71

This analogy is not intended to suggest that the causative role of morphogenetic fields depends on conscious design, but only to emphasize that not all causation need be energetic, even though all processes of change involve energy. The plan of a house is not in itself a type of energy./71

Motor fields, like morphogenetic fields, are given by morphic resonance from previous similar systems. The detailed structure of an animal and the patterns of oscillatory activity within its nervous system will generally resemble itself more closely than any other animal. Thus the most specific morphic resonance acting upon it will be that from its own past./170

This is an impossible book from an impossible author. This not being enough, it's besides also a book that really angered me. Never have I read such gimmick from an author who doesn't even has to courage to stand up for his materialistic bias, but hides behind a good boy façade of elegant verbiage to bring over a message that basically is none! One of the critics, while he was really off-track in his arguments against Sheldrake's research, wrote that the worst a scientist could do in his life was to write a book that was a waste of time to read for other scientists. And I think that this is really true, and that that was ultimately the reason of my anger.

Published by Sirius-C Media Galaxy LLC, 2011

RICK STRASSMAN

DMT – The Spirit Molecule

A Doctor's Revolutionary Research into the Biology of Near-Death and
Mystical Experiences
Rochester: Park Street Press, 2001

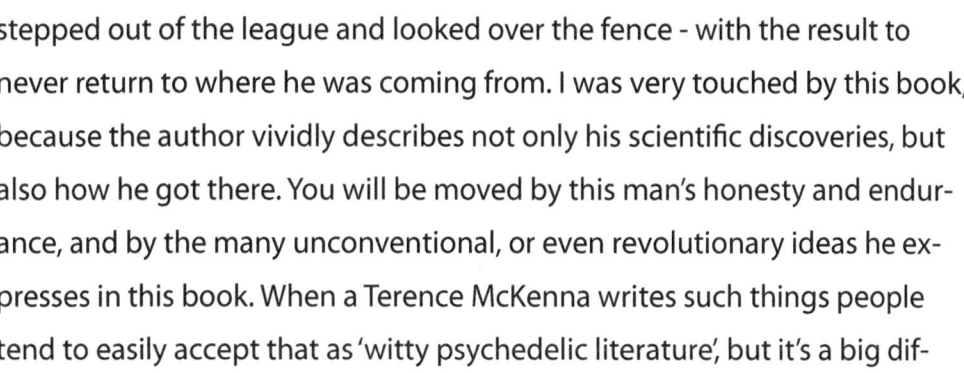

DMT – The Spirit Molecule is a courageous book of
a remarkable American doctor, one who really
stepped out of the league and looked over the fence - with the result to
never return to where he was coming from. I was very touched by this book,
because the author vividly describes not only his scientific discoveries, but
also how he got there. You will be moved by this man's honesty and endur-
ance, and by the many unconventional, or even revolutionary ideas he expresses in this book. When a Terence McKenna writes such things people
tend to easily accept that as 'witty psychedelic literature', but it's a big difference when a medical doctor writes about

what is considered by many as *tabooed research*,
taking the obvious risk to be violently discarded
out of the peer group. We know from the past
how that can happen. The life stories of, for example, Dr. Franz Anton Mesmer and Dr. Wilhelm
Reich give a vivid and picturesque account of it.
But Strassman does not want to get out a book that goes in the face of
some people. He is not that kind of character. I got the impression throughout this book that he is a mature personality and knows what he is talking
about. So much the more need we to respect this voice of authority in a
jungle of information about a worldview that I would call *centered worldview*, as opposed to the *scattered worldview* that is the day-to-day condition
of modern consumer culture.

In my view, Strassman is more outspoken than for example Stan Grof when it goes to clearly state the disaster that was done by governmental authority to prohibit LSD and a whole array of powerful entheogens that were used, with good care, in experimental psychiatry for finding a better, and more effective, approach to healing the mind of the whole human. It needs courage to pronounce heretic views of this kind from the pulpit of an accredited doctor, because it can result in professional ruin. That this man has taken the courage to walk his talk despite the risk needs a big applause, from what community or point of view we look at him! To begin with, he writes as a general introduction on the subject of psychedelics:

Dr. Rick Strassman

Psychedelics were the growth area in psychiatry for over twenty years. Now young physicians and psychiatrists know nearly nothing about them./27

On the other hand, some progressive movements who foster abortion rights may be disappointed about his total reject of abortion, a view that by the way I myself personally support since many years. He gives conclusive evidence for the point he makes, and if this evidence will be corroborated by further research, I am quite certain that legally sanctioned abortion will be abandoned in the future. It's quite a hot issue, but unfortunately the debate only focuses on the rights of the mother. What about the rights of the fetus to be born, given that he or she, if the mother was raped or not, has decided to incarnate? When this is already sound from a spiritual point of view, the scientific evidence that Strassman gives for his stance on rejecting abortion speaks the same language:

Dr. Rick Strassman

Opponents and supporters of abortion rights may find fault with my proposal that a pineal DMT release at forty-nine days after conception marks the entrance of the spirit into the fetus./XVII

If we are to respect life at all, we have to respect it from the moment it's *animated* matter, which is matter that serves as a vehicle for spirit. This is recognized in all but Western cultures as one of the base principles of life, and I can't see why we should make an exception for abortion.

Now, to proceed in this book review, I cannot produce all the quotes from the book that I have taken when reading it, because of copyright reasons. So let me just provide a few references, and comment on them. In his concise overview over the history of using psychedelic compounds in psychiatry, Dr. Strassman gives much food for thought that supports the alternative position, the one taken, for example, by Terence McKenna, Richard E. Schultes, Jeremy Narby, Ralph Metzner, or Stan Grof, to name only a few[16]:

> **Dr. Rick Strassman**
>
> Psychedelic research was a bruising and humiliating chapter in the lives of many of its most prominent scientists. These were the best and the brightest psychiatrists of their generation. Many of today's most respected North American and European psychiatric researchers, in both academics and industry, now chairmen of major university departments and presidents of national psychiatric organizations, began their professional lives investigating psychedelic drugs. The most powerful members of their profession discovered that science, data, and reason were incapable of defending their research against the enactment of repressive laws fueled by opinion, emotion, and the media./28

It also was shown by research that LSD has a powerful effect on enhancing creativity:

> **Dr. Rick Strassman**
>
> The late Willis Harman possessed one of the most discerning minds to apply itself to the field of psychedelic research. Earlier

---

[16] See also Pierre F. Walter, *The New Paradigm in Consciousness, Psychology, Healing and Spirituality, Book Reviews (2011)*.

in his career, he and his colleagues administered LSD to scientists in an attempt to bolster their problem-solving skills. They found that LSD demonstrated a powerfully beneficial effect on creativity./XVI

Now, it is quite obvious that it is not a very comfortable condition for a scientist to do research in an area and on a subject matter that is a potential case of taboo, because the law suddenly shifted and declared the specific topic of research an illegal matter. But let us see what that means? Is science restricted to research only in matters that are legal? Is scientific curiosity limited to what the law givers think and enact? Apart from the ethical foundation of science, that by the way was never really questioned in the governmentally funded research on genetic manipulation and technology, and where there are real dangers, not just dangers in the heads of paranoid politicians, science should in my view not be restricted to what is declared legal, but overall needs to serve the *progress of humankind*.

This is my own position as an international lawyer on this subject. Dr. Strassman makes a good additional point stating that the mere absence of academic attention for any given subject of research should not keep curious scientists from investigating in the matter to find out what is true, and what is myth:

Dr. Rick Strassman
The lack of academic attention to psychedelics may have been partly due to the absence of any ongoing human research. However, it is common for physicians-in-training to learn about previously popular theories and techniques, even if they no longer are in favor. The psychedelic drugs, however, seemed to have dropped out of all psychiatric dialogue./28

Published by Sirius-C Media Galaxy LLC, 2011

A matter of research that has hardly been tackled yet by modern science is DMT-induced trance that is brought about by the human body's own DMT production. Dr. Strassman, based on a lot of former research that he cites in the notes, is convinced that the pineal gland produces endogenous DMT. It is true that psychedelics explorers such as Terence McKenna have stressed the fact that DMT has strong affinities with the human mindbody in that it's a compound that the human body produces itself, while this is, for example, not the case for LSD. Dr. Strassman notes:

Dr. Rick Strassman

The similarities between naturally occurring and DMT-induced phenomena support my suggestion that spontaneously occurring 'psychedelic' experiences are mediated by elevated levels of endogenous DMT./311

Eventually, as this book review must remain eclectic because of the complexity of the author's research that can only be comprehended when you actually read the book, I would like to stress another quite unusual characteristic of the author. He is a spiritually initiated person, not the usual vintage of your Cartesian-freak physician. I mention this because you can draw from this book much insight about the modern-day connection between science and spirituality, and between medical science and spirituality. Let me give you, to come to the end of this review, the following quote:

Dr. Rick Strassman

In Judaism, for example, consciousness moves through the sefirot, or Kabbalistic centers of spiritual development, the highest being Keter, or Crown. In the Eastern Ayurvedic tradition, these centers are called chakras, and particular experiences likewise accompany the movement of energy through them. The highest chakra is also called the Crown, or the Thousand-Petaled Lotus. In both traditions, the location of this Crown sefira or chakra / is

the center and top of the skull, anatomically corresponding to
the human pineal gland./58-59

To summarize, this book is highly recommended for the serious researcher
and the non-scientist interested in revolutionary new insights from cutting-
edge research that sooner or later will lead to an explosion of mainstream
medical science and psychology because of a paradigm shift so gigantic,
and so irreversible, that it will have an impact on all of us. And the doctor's
profession will probably never be as it was before.

Dr. Strassman has great merit to have the courage and the great synthetic
intelligence to present this research to us in a book that was for him much
more than a book. As he gives much personal information in the book, this
will become obvious for you as well once you read it. And it might fill you
with awe, and you might think that you can meet heroes also outside of the
fiction-world of television, and in real life!

MICHAEL TALBOT

The Holographic Universe

New York: Harper Perennial, 1992

The Holographic Universe by **Michael Talbot** is an extraordinary book, and a captivating read from the first to the last page. Not only has this book merited more than five stars, it merits both a literary prize and a distinction for exemplary scientific research. As the author has already passed over into greater life dimensions, these are posthumous laurels. But my recognition of the author's genius, his motivation, his purity, and his literary and spiritual maturity is real. The book has been a true companion for me, and it accompanied me on all trips over weeks.

To begin with, the author has not just delivered an extensive study on the vast realm of psychic phenomena, he also, similar to Fritjof Capra, has consulted personally a number of researchers he refers in his book. Myself a researcher, I know well that there is a difference between pure book research, or to actually interview researchers and authors, and thus gain direct and personal insight into the human experience that is behind the book or research.

Honestly, I have seldom read a study that was both so well-researched and well-written as this book. As Talbot revealed in a few notes, and rather modestly, he had himself strong psychic abilities and was a psychic already as a child. This may explain in part his participatory experience as a scientist and his fundamental comprehension of the topics at stake. In fact, he comes over as an expert on the topics at stake, while he does not boast with his knowledge or his paranormal abilities.

Talbot makes a strong point for the holographic nature of the universe and of psychic experiences in general, and I can virtually not see how his theory can be refuted, so well-founded it appears. An important point of departure, just as in Lynne McTaggart's startling book *The Field (2002),* is the acknowledgment of living space where former Cartesian science has claimed to find a vacuum or *empty space*:

Michael Talbot

Space is not empty. It is full, a plenum as opposed to a vacuum, and is the ground for the existence of everything, including ourselves./51

It just makes sense of so much material that at first sight seems scattered or even completely inconsistent, such as, for example, UFO sightings and alien abduction experiences. Under the header of the holographic view of the universe, not only such experiences, but also so apparently different subjects as Jewish Cabala and David Bohm's *implicate order* interpretation of quantum physics perfectly and intelligently correspond to each other. While I recommend this book without hesitation for the general reader, it should be an advantage to have read previously some kind of fundamental introduction in the new paradigms in science, such as, for example, Fritjof Capra's *The Turning Point (1982/1987)*. Talbot writes:

Michael Talbot

In his general theory of relativity Einstein astounded the world when he said that space and time are not separate entities, but are smoothly linked and part of a larger whole he called the space-time continuum. Bohm takes this idea a giant step further. He says that everything in the universe is part of a continuum. Despite the apparent separateness of things at the explicate level, everything is a seamless extension of everything else, and ultimately even the implicate and explicate orders blend into each other./48

What I normally not do, I shall do it here: reproduce the *Table of Contents* including the sub-chapters. This will by itself give an overview over the vast array of topics and research that the author has summarized in this extraordinary book, the Notes section of which expands over twenty-four pages! Below I will do a pointed discussion based on some quotes from the book.

PART ONE - A REMARKABLE NEW VIEW OF REALITY

1 The Brain as Hologram

2 The Cosmos as Hologram

PART TWO - MIND AND BODY

Published by Sirius-C Media Galaxy LLC, 2011

I have read the book once but there are so many details from so many areas of research, that I have begun reading selected chapters again, and I vividly recommend this to the reader. I have collected many quotes of the book but will present only a few, and only on synthetic insights, where the author summarizes specific fields of research in order to make the point for a valid *holographic paradigm* to currently emerge in global science.

To begin with, when we try to summarize the most important insight from quantum physics, we could describe it with the word *participatory*; this is not just Talbot's personal view, but a shared assumption about quantum physics. For example, Lynne McTaggart says the same in her book *The Field*. Talbot writes:

Michael Talbot

As we have seen, one of the basic tenets of quantum physics is that we are not discovering reality, but participating in its crea-tion. It may be that as we probe deeper into the levels of reality beyond the atom, the levels where the subtle energies of the

Published by Sirius-C Media Galaxy LLC, 2011

human aura appear to lie, the participatory nature of reality becomes even more pronounced. Thus we must be extremely cau-
tious about saying that we have discovered a particular structure
or pattern in the human energy field, when we may have actually
created what we have found./191

The author lucidly adds the note here that with this change of the basic science paradigm from an observatory to a participatory experimental setup
of the scientific task, the role of the scientist changes implicitly:

Michael Talbot
A shift from objectivity to participation will also most assuredly
affect the role of the scientist. As it becomes increasingly apparent that it is the experience of observing that is important, and
not just the act of observation, it is logical to assume that scientists in turn will see themselves less and less as observers and
more and more as experiencers. As Harman states, 'A willing-
ness to be transformed is an essential characteristic of the participatory scientist./298

The next important characteristic of a holographic universe that we can derive from quantum physics is the so-called *nonlocality principle*. Talbot
writes:

Michael Talbot
Just as an image of an apple has no specific location on a piece
of holographic film, in a universe that is organized holographi-
cally things and objects also possess no definite location; everything is ultimately nonlocal, including consciousness./234

Another important part of Talbot's research were *Near Death Experiences
(NDEs), Out-of-Body Experience (OBEs)* and voyages into past lives through
past-life regression therapy and in-between-lives regression. The author
summarizes the research results as follows:

Michael Talbot

Several said they didn't even have a body unless they were thinking. (...) But as their experience in the between-life state continued, they gradually became a kind of hologramlike composite of all of their past lives. This composite identity even had a name separate from any of the names they had used in their physical incarnations, although none of his subjects was able to pronounce it using their physical vocal cords. (...) Many say that they were not aware of any form and were simply 'themselves' or 'their mind'. Others have more specific impressions and describe themselves as 'a cloud of colors', 'a mist', 'an energy pattern' or 'an energy field', terms that / again suggest that we are all ultimately just frequency phenomena, patterns of some unknown vibratory energy enfolded in the greater matrix of the frequency domain./247-248

Some NDEers assert that in addition to being composed of colored frequencies of light, we are also constituted out of sound. 'I realized that each person and thing has its own musical tone range as well as its own color range', says an Arizona housewife who had an NDE during childbirth. 'If you can imagine yourself effortlessly moving in and out among prismatic rays of light and hearing each person's musical notes join and harmonize with your own when you touch or pass them, you would have some idea of the unseen world. The woman, who encountered many individuals in the afterlife realm who manifested only as clouds of colors and sound, believes the mellifluous tones each soul emanates are what people are describing when they say they hear beautiful music in the ND-dimension.'/248

Talbot did not only research in present-time science theories and experimental research, but also looked at what spiritualistic authors wrote, as for example Emanuel Swedenborg. And he discovers evidence for the holographic theory in Swedenborg's writings:

Published by Sirius-C Media Galaxy LLC, 2011

Michael Talbot

Most intriguing of all are those remarks by Swedenborg that seem / to refer to reality's holographic qualities. For instance, he said that although human beings appear to be separate from one another, we are all connected in a cosmic unity. Moreover, each of us is a heaven in miniature, and every person, indeed the entire physical universe, is a microcosm of a greater divine reality./258-259

Swedenborg also believed that, despite its ghostlike and ephemeral qualities, heaven is actually a more fundamental level of reality than our own physical world. It is, he said, the archetypal source from which all earthly forms originate, and to which all forms return, a concept not too dissimilar from Bohm's idea of the implicate and explicate orders./259

Another important area of research into holographic phenomena in Talbot's book are visions of the Holy Virgin, and angels. Talbot writes:

Michael Talbot

Another impressively holographic Marian vision is the equally famous appearance of the Virgin in Zeitoun, Egypt. The sightings began in 1968 when two Moslem automobile mechanics saw a luminous apparition of Mary standing on the ledge of the central dome of a Coptic church in the poor Cairo suburb. (...) Most telling of all, after three years of manifestations and as interest in the phenomenon started to wane, the Zeitoun figures also waned, becoming hazier and hazier until, in their last several appearances, they were little more than clouds of luminous fog./275

Other areas of research that I mention here eclectically, as it's virtually impossible to comment on all research topics, are the *Tibetan Book of the Dead* and the spiritual teachings of native peoples. The author writes:

Michael Talbot

One thing that we do know is that in a holographic universe, a universe in which separateness ceases to exist and the innermost processes of the psyche can spill over and become as much a part of the objective landscape as the flowers and the trees, reality itself becomes little more than a mass shared dream. In the higher dimensions of existence, these dreamlike aspects become even more apparent, and indeed numerous traditions have commented on this fact. The Tibetan Book of the Dead repeatedly stresses the dreamlike nature of the afterlife realm, and this is also, of course, why the Australian aborigines refer to it as the dreamtime. Once we accept the notion, that reality at all levels is omnijective and has the same ontological status as a dream, the question becomes, Whose dream is it?/285

Like Bohm, who says that consciousness always has its source in the implicate, the aborigines believe that the true source of the mind is in the transcendent reality of the dreamtime. Normal people do not realize this and believe that their consciousness is in their bodies. However, shamans know this is not true, and that is why they are able to make contact with the subtler levels of reality./289

Eventually, the author finds examples from the past where scientists were already looking beyond the fence and perceiving the universe as basically interconnected and *nonlocal*, as we would say today. And he mentions the German mathematician Leibniz who found that the I Ching was coded in a binary way, as we today know computers work.

Michael Talbot

In short, long before the invention of the hologram, numerous thinkers had already glimpsed the nonlocal organization of the universe and had arrived at their own unique ways to express this insight. It is worth noting that these attempts, crude as they may seem to those of us who are more technologically sophisti-

Published by Sirius-C Media Galaxy LLC, 2011

cated, may have been far more important than we realize. For instance, it appears that the seventeenth-century German mathematician and philosopher Leibniz was familiar with the Hua-yen school of Buddhist thought. Some have argued that this was why he proposed that the universe is constituted out of fundamental entities he called 'monads', each of which contains a reflection of the whole universe. What is significant is that Leibniz also gave the world integral calculus, and it was integral calculus that enabled Dennis Gabor to invest the hologram./291

I will come to an end here with this rather extensive book review, and hope that it conveys the magnificence of the author's work and research, and his ability to convey the essential of it in clear language, including a whole array of research done by other scientists that he mentions in the huge Notes section.

I highly recommend this book, also to the non-initiated and non-scientific reader. The author somehow manages to express himself with common sense and insight, without using complex scientific language. This book can also be taken as a useful research library as from the Notes you can stretch out your research further, about every topic of research.

WHAT THE BLEEP DO WE KNOW!?

DVD Set

Producers: William Arntz, Betsy Chasse
Directors: William Arntz, Marc Vincente, Betsy Chasse
Sceenwriters: William Arntz, Betsy Chasse
Twentieth Century Fox LLC
DVD Set, 360 min.

I shall discuss both volumes here in one review, and this for good reason. *The Rabbit Hole Edition* or *Quantum Edition* is for the most part a summary and expansion of the *Bleep* - with countless scenes repeated from the first edition, but cut in a different way. In effect, the cut is highly interesting; it seems to me that part of the convincing magic of these films is the cut. This is especially true for the *Quantum Edition*, 3rd DVD.

Each of the three DVDs contains elements of the *Bleep*, scenes that are repeated in part, not in full, so as to fuse them together with new elements. This is highly educative, and it seems to me that indeed, the intention of the producers here was one of explication, one of condensation and one of still more detailed explanation of the main theme of the whole business: creating our own reality and what it means. The little critter, as it were, is set out most in detail in the 3rd and last DVD of the Quantum Edition, which is in my opinion the best and most genuine of all.

Dr. Fred Alan Wolf

The basic message of the extended movie is that science is eminently spiritual and holistic, and cannot be in our days be defined as Cartesian as it was during the past two hundred fifty years. These scientists and authors interviewed in the movie unanimously vote for a *New Science* that is beyond reductionism and human in that it encompasses the spiritual dimension of the human being in every sense of the word, and in that it is systemic and shows the hidden connections of life and living, within the greater metaverse. I regret only that Ervin Laszlo was not interviewed for the movie!

Thereby, the movie actually goes one step ahead, showing that science is only one tool of many, and that the other essential tool is awareness building, the developing of human consciousness as a primary tool to gain an objectivized view of reality that is beyond personal frenzies and emotional addiction. The merit of the movie is to exhibit scientific and complexity of living in a way even intelligent lay people, policy makers and non-scientists can understand.

For my comprehension level, nothing is left open, except the essential truth that creating your reality does not mean you are going to be happy and rewarded for it.

I think that most people don't even get through to this insight, because of their Cartesian doubts and their reductionist mindset. They doubt the whole process and that's why, I guess, so much is repeated over and over in the *Quantum Edition*.

In fact, I work since now fifteen years with the scientific prayer method that is an integral part of *Divine Science*, as I write more about it on Science and Divination and in my reviews of some of Dr. Murphy's books. When you consider that this method was first invented by Ernest Holmes in 1927, and expanded and commercialized in the 1960s, by Dr. Joseph Murphy and Catherine Ponder, then you are not actually so surprised about the Bleep and Rabbit Hole films. They tell you a familiar reality. What I am saying is thus that they don't tell something entirely new, but they tell it in a very good, pedagogically sound and, as it were, educational way. And they back this truth up with a lot of scientific data that confirms that we are indeed the creators of our reality, and co-creators with the Divine. And all this is done in a way that is also aesthetically very beautiful, more beautiful than anything I have seen of this kind before. It's a wonderful fusion of truth with art, if I may say so.

Dr. William A. Tiller

However, there is more to creating your own reality, and what I am going to say now was not said in any way in the movie. Why? Because it's not so wonderful after all, and not so positive. But life is not always positive, as movies are. This secret is that when you have created a reality, be it art or whatever you manifest as reality in your life, that does not mean that others will recognize it as such, it does not even mean that they are going to notice it, let alone appreciate it. It doesn't mean you will be rewarded for it. It doesn't mean that you are going to 'make money' with it. It can mean that you are still more alone than before, still more isolated and still more re-jected …, so the outcome is not entirely predictable. This, of course, is not said in the film, and that is what it molds it into what it is and was intended to become: a commercial product. Not a new Bible. So definitely, you should let a chamber open in your inquisitive mind and stay with one ques-tion:

– What the Bleep does The Bleep know?

WILLIAM A. TILLER

Conscious Acts of Creation

Producer: Associated Producers, Inc.
DVD, 90 min.

'In *Conscious Acts of Creation*, Dr. Tiller makes a claim that would not only revolutionize medicine but our perception and approach to all reality. Unlike many who venture into these realms, Tiller has a distinguished career at Stanford University and a solid grounding in physics. If there are prophets in our extraordinary times, he is likely one of them.'
- Dr. Wayne B. Jonas, MD

Former Director of the Medical Research Fellowship at Walter Reed and the Office of Alternative Medicines at the National Institutes of Health, and author of over 60 publications.

Based upon years of detailed research, Dr. Tiller has amassed convincing experimental data showing that in seemingly the same cognitive space, basic chemical reactions and basic material properties can be strongly altered by human intentions. Essentially, he says, we are all capable of performing what we typically think of as miracles.

In this presentation, Dr. Tiller explains these findings in clear, understandable language and supports it with just enough math and physics to deeply move just about everyone. The exciting bottom line: It appears that these findings ... and the new technologies he has been working with, are capable of advancing every single industry that humanity is presently engaged in, and can perhaps even create new ones!
From: DVD Back Cover

Conscious Acts of Creation is not a DVD in the strict sense of the word, but a scientific presentation. And one of the purest and driest kind. No animations, no movies, no fun. There is a small and probably highly erudite audience, and a scientist who speaks in front of it for 90 minutes.

And he shows a number of overheads, and talks a lot mathematical and physical stuff, including equations that I whole-heartedly congratulate you to understand. For me, unfortunately, it was a number too high, and that's not Tiller's fault at all. I was not on the best terms with our physics teacher at high school, called him a Cartesian freak and dared to question Einstein's relativity theory.

This being said, I do not say that Tiller speaks gibberish. He is able, very able even, to explain all this in very understandable language. But he had to give some flesh and bones, of course, for those 'from the faculty' and for those who are going to turn his theory upside down so often and so thoroughly that it will no more be flesh and bones but *hack meat*. The good news is that he himself has done that before them, in a sense of wise precaution, so they won't get very far tearing him down into the gutter, as they did with so many before him.

He is a very careful man and scientist and he's not the person to boast up with something light-heartedly. Also when you consider his age and the huge reputation he could lose, you can be almost sure that this man speaks from an inner Kantian imperative - because he has to convey a very important message to humanity. To me, Dr. Tiller comes over as plainly honest, a man who walks his talk and is one of the pioneers of *New Science*.

What is his research about? It's about the most daring subject I ever came across since I do research on new science and quantum physics. It's how human intention or intentionality changes the very structure of matter, the very molecules that matter consists of, the very form and coding of matter. I believe that still two decades ago, he would have been dismissed as a delusionist, but we are definitely a big step ahead in the meantime in defining the new paradigm of science, and Tiller is, next to Reich, Burr, Capra, Laszlo, Talbot, McTaggart, Radin, Goswami, Pert, Sheldrake and a few others an avatar who predicts a large and brilliant future of spiritual science in every sense of the word. His research is convincing and backed up by hard physics and mathematical data that are simply on the ta-  ble – hard to refute. Randi and their likes have lost the game, while I do not contest that their skepticism has contributed to unveil some *New Age* scam artists! Tiller is beyond that, because he is beyond fake science. He is perhaps making history in being one of the precursors of the real science of the future.

Published by Sirius-C Media Galaxy LLC, 2011

VIDETTE TODARO-FRANCESCHI

The Enigma of Energy

Where Science and Religion Converge
New York: Crossroad, 1999

The Enigma of Energy by Vidette Todaro-Franceschi is a carefully researched study, originally a PhD thesis, that treats a very unusual subject, the cosmic life energy. The author has fulfilled a Sisyphus task with this seminal work that represents a remarkable scientific achievement. It's not an easy-read, but for the serious researcher, it's an invaluable resource. It has been for me.

The author takes a somewhat unusual position right from the start of her daring research, conscious that it would be a multi-disciplinary study. And she reports that her study took her much farther than she had believed at first, and that the deeper she researched into the cosmic energy, the more she found a connection of her research with religion. She writes in the Introduction:

> **Vidette Todaro-Franceschi**
> The more I worked on this project the more I became aware that somehow science and religion were converging. It was never my goal to merge these two seemingly disparate areas; in fact, when my search led me into religious realms of thought, I tried hard at first to stay clear of them. But it was impossible to do so. Anytime I came across literature that was related to an idea of energy there were implicit or explicit spiritual overtones. Most surprising was the abundance of spiritual ideas found in physics./4

With her wide and strongly intuitive vision, the author approaches the topic in a very systematic and methodologically sound manner without getting lost in the maze of a thousand and one philosophical concepts that express with different terms what is without a doubt one and the same thing. She first looks at the etymology of the word *energy*, and the concepts of energy in various cultures and with different philosophical traditions. And she writes:

Vidette Todaro-Franceschi

To begin at the beginning is to begin with Aristotle's conceptualization of energeia./4

While taking her starting point with Aristotle, the author also looked beyond the fence of Western tradition and into the very explicit Asian notions of the bioenergy. Under the header *Ideas of Energy in Antiquity*, she writes:

Vidette Todaro-Franceschi

In the East, the ancient Chinese held that the universe was a dynamic entity filled with continuous cyclic flow and change. The motion of these cyclic patterns is expressed by the concepts of yin and yang. The yin is the female, dark, quiet and resting, intuitive component of change and is associated with the earth. The yang is the male, light, strong, creative, active component of change and is associated with heaven. Although polar opposites, together the yin and yang comprise life, where there is a continual harmonizing of both cycles of change. These two concepts indicate the underlying Tao (way) or pattern of everything in the universe. The inherently changing nature / of things is held in balance or harmony by a continuous flow of ch'i./13-14

Prana, a term that has been used in the ancient Indian tradition for over five thousand years, denotes a universal or vital life energy. Often translated as 'breath' in Indian works, the concept of prana is a central one for principles of healing in Indian Ayurvedic medicine, which was developed from 1200 to 800 B.C. Today adherents believe that one can regulate and manipulate the flow of prana in order to maintain or restore health using various alternative health care modalities, such as healing touch./14

But the focus of the work is clearly more on the side of the Western notion or notions of energy, than on the side of the cosmic life energy, the one energy that is behind all manifestations and the visible reality. And for this purpose, and with this focus, the dissertation has great merit.

Being myself a bioenergy researcher, and having let my research start out with Paracelsus, I am conscious that I left out Aristotelian speculations; for that's what they are. I have had a look at Aristotle's *energeia* idea before reading her study, but I did not find a clear line, nor applicable result of Aristotle's considerations. In fact, I found his ideas about energy interesting, but entirely intuitive. They may well one day prove to be correct, but I can't see how they can be evaluated as being scientific. The author managed to very quaintly summarize the Aristotelian teaching about energy, and this is one of the greatest merits of her concise study on the conceptual framework of energy over the whole of human history:

Vidette Todaro-Franceschi

In The Metaphysics (Book IX, Theta) Aristotle noted the importance of energeia in relation to dynamis (potentiality) and kinesis (change or motion). He distinguishes two types of energeia. The first type of energeia is an unended or imperfect actuality, for example, walking or building. These energeiai are said to be movements or kinesis, because they are incomplete. The second type of energeia is a complete or perfect actuality, an entelecheia, where the end product is within the thing itself, for example, seeing and thinking. These energeiai are referred to in general as actualities./20

Toward the end of her study, the author asks the question 'What, Exactly, is Nature?' Referring to historian and philosopher of science R. G. Collingwood, she writes that there are three periods in the development of the idea of *nature*, which she sees coincidentally reflect the ideas of energy.

Vidette Todaro-Franceschi

In his discussion of the first period, the Greek view of nature, Collingwood points out that the ancient Greeks believed a certain vitality or ceaseless motion existed in nature, which they generally attributed to the soul. (...) The most important aspect of Aristotle's conception of nature lies in his belief that all things have a final cause, which is exhibited by the individual thing's form. According to him the soul was the essence of living things, and of course the form of anything / was the purpose or reason for its becoming. Overall, according to Aristotle, the teleological qualities of things were so strong that there could be no explanation for anything in nature, including us, without it./123-124

In the second period, she reports, that Collingwood referred to as the Renaissance view of nature, the mechanistic worldview was firmly established.

Vidette Todaro-Franceschi

Collingwood notes that the second stage of the Renaissance view of nature came about with the Copernican discovery that our world was not the center of the universe. The main contention during this time became 'the denial that the world of nature, the world studied by physical science, is an organism and the assertion that it is both devoid of intelligence and of life'. / During this period, human beings were seen as outside of, rather than a part of, nature. We became pompous, thinking that we controlled things and that we were somehow superior. Explicit in this view was the denial of final causation. The primary focus was on matter and the natural laws by which matter changes. Science and philosophy recognized only efficient causes: forces producing effects. And finally, mathematical structure accounted

Published by Sirius-C Media Galaxy LLC, 2011

for the changes, both of a qualitative and quantitative nature./124-125

The third and last period, she reports that was identified by Collingwood was the modern view of nature, which had its origin in the latter part of the eighteenth century when process and change became the focus.

Vidette Todaro-Franceschi

I believe that during this period the idea that energy was an autonomous existent contributed to the shift in focus. It became vaguely evident that change was inherent in various things; that is, it was recognized that change could occur without the provocation of external forces or efficient causes./125

Collingwood identifies the idea of a 'rhythmical pattern' with the modern view of nature and acknowledges that the new physics theories are partly responsible for this notion. But the rhythmical patterns we now know to exist in nature also seem to denote an inner principle of change, or an Aristotelian 'that for the sake of which', originally expressed by the ancient Greeks. So one might say we have come full circle.

In conjunction with this new take on an old idea that was present in both Eastern and Western antiquity is the increasing awareness that intuition plays a significant role in scientific discoveries. As the historical background of the idea of energy attests, intuitive ways of knowing have been crucial to the development of scientific ideas throughout history. Many individuals knew things, such as the energy conservation doctrine, without being able to empirically verify them. In other words, intuitive ways of knowing seem to have led / us in the right direction long before we were capable of scientifically validating what we somehow knew to be so.

> Subjectivity and subjective ways of knowing, such as intuition, have become as vital to our understanding as objectivity and empirical ways of knowing. In this modern view of nature humankind has once again come to be recognized as being part of nature, rather than outside of it./125-126

I will come to an end of my book review here by not overdoing to quote from this interesting study that I fully recommend and endorse as one of the best books written so far on the historical and philosophical development of the concept of energy, and the implications of it for the progress of science, and its opening to the higher dimension of understanding the meaning of life, and thus its implicit connection with religion. The study provides ample references for the researcher and its structure is scientifically meticulous, and methodologically convincing.

I would have wished inclusion of the research of Paracelsus, Mesmer, Reich, Reichenbach, Burr or Lakhovsky on the cosmic energy field, the consideration of Ayurveda, Homeopathy and especially Bach Flowers and their amazing vital energy resources, and the ample conclusions we can derive from the old Chinese science of Feng Shui, but this, then, shall be the focus of further research, and is actually the main focus of my own research on the cosmic life energy.

On the other hand, the study provides pertinent information on the roots of the energy doctrine in our own Western philosophical and science tradition, and here is its intrinsic value. I have not read anything of this quality so far about Aristotle's insights into the systemic properties of nature, and I have greatly profited from the knowledge and deep understanding that this study provides for Aristotle's doctrine of *energeia*. As this is not easy to understand a matter by reading the originals, I have come to greatly appreciate the intellectual work done by the author, and the penetration of the matter she achieved, and this will certainly be awarded one day.

Published by Sirius-C Media Galaxy LLC, 2011

BIBLIOGRAPHY

Contextual Bibliography

Arntz, William & Chasse, Betsy

What the Bleep Do We Know
20th Century Fox, 2005 (DVD)

Down The Rabbit Hole Quantum Edition
20th Century Fox, 2006 (3 DVD Set)

Bleep
An der Schnittstelle von Spiritualität und Wissenschaft
Verblüffende Erkenntnisse und Anstösse zum Weiterdenken
Berlin: Vak Verlag, 2007

Bertalanffy, Ludwig von

General Systems Theory
Foundations, Development, Applications
New York: George Brazilier Publishing, 1976

Boadalla, David

Wilhelm Reich, Leben und Werk

Bohm, David

Wholeness and the Implicate Order
London: Routledge, 2002

Die implizite Ordnung
Grundlagen eines dynamischen Holismus
München: Goldmann Wilhelm, 1989

Thought as a System
London: Routledge, 1994

Quantum Theory
London: Dover Publications, 1989

La plénitude de l'univers
Paris: Rocher, 1992

La conscience de l'univers
Paris: Rocher, 1992

Die unbekannten Schriften der Essener
Saarbrücken: Neue Erde/Lentz, 2002

Brennan, Barbara Ann

Hands of Healing
A Guide to Healing Through the Human Energy Field
New York: Bantam, 1988

Published by Sirius-C Media Galaxy LLC, 2011

Bruce, Alexandra

Beyond the Bleep
The Definite Unauthorized Guide to 'What the Bleep Do we Know!?'
New York: Disinformation, 2005

Burwick, Frederick

The Damnation of Newton
Goethe's Color Theory and Romantic Perception
New York: Walter de Gruyter, 1986

Campbell, Herbert James

The Pleasure Areas
London: Eyre Methuen Ltd., 1973

Der Irrtum mit der Seele
München: Scherz Verlag, 1973

Les principes du plaisir
Paris: Stock, 1974

Capra, Bernt Amadeus

Mindwalk
A Film for Passionate Thinkers
Based Upon Fritjof Capra's *The Turning Point*
New York: Triton Pictures, 1990

Capra, Fritjof

The Turning Point
Science, Society And The Rising Culture
New York: Simon & Schuster, 1987
Original Author Copyright, 1982

Wendezeit
Bausteine für ein neues Weltbild
München: Droemer Knaur, 2004

Le temps du changement
Science, société et nouvelle culture
Paris: Rocher, 1994

The Tao of Physics
An Exploration of the Parallels Between Modern
Physics and Eastern Mysticism
New York: Shambhala Publications, 2000
(New Edition) Originally published in 1975

Das Tao der Physik
Die Konvergenz von westlicher Wissenschaft und östlicher Philosophie
Neue und erweiterte Auflage
München: O.W. Barth bei Scherz, 2000

Ursprünglich erschienen 1975 bei Droemersche Verlagsanstalt in Hamburg

Le tao de la physique
Paris: Sand & Tchou, 1994

The Web of Life
A New Scientific Understanding of Living Systems
New York: Doubleday, 1997
Author Copyright 1996

Lebensnetz
Ein neues Verständnis der lebendigen Welt
München: Scherz Verlag, 1999

The Hidden Connections
Integrating The Biological, Cognitive And Social
Dimensions Of Life Into A Science Of Sustainability
New York: Doubleday, 2002

Verborgene Zusammenhänge
München: Scherz, 2002

Steering Business Toward Sustainability
New York: United Nations University Press, 1995

Uncommon Wisdom
Conversations with Remarkable People
New York: Bantam, 1989

The Science of Leonardo
Inside the Mind of the Great Genius of the Renaissance
New York: Anchor Books, 2008
New York: Bantam Doubleday, 2007 (First Publishing)

Complete List of Publications
http://www.fritjofcapra.net/publishers.html

Cayce, Edgar
Modern Prophet
Four Complete Books
'Edgar Cayce On Prophecy'
'Edgar Cayce On Religion and Psychic Experience'
'Edgar Cayce On Mysteries of the Mind'
'Edgar Cayce On Reincarnation'
By Mary Ellen Carter
Ed. by Hugh Lynn Cayce
New York: Random House, 1968

Published by Sirius-C Media Galaxy LLC, 2011

Cho, Susanne

Kindheit und Sexualität im Wandel der Kulturgeschichte
Eine Studie zur Bedeutung der kindlichen Sexualität unter besonderer
Berücksichtigung des 17. und 20. Jahrhunderts
Zürich, 1983 (Doctoral thesis)

Chopra, Deepak

Creating Affluence
The A-to-Z Steps to a Richer Life
New York: Amber-Allen Publishing (2003)

Life After Death
The Book of Answers
London: Rider, 2006

Leben nach dem Tod
Das letzte Geheimnis unserer Existenz
Berlin: Allegria Verlag, 2008

Synchrodestiny
Discover the Power of Meaningful Coincidence to Manifest Abundance
Audio Book / CD
Niles, IL: Nightingale-Conant, 2006

The Seven Spiritual Laws of Success
A Practical Guide to the Fulfillment of Your Dreams
Audio Book / CD
New York: Amber-Allen Publishing (2002)

Die Sieben Geistigen Gesetze des Erfolgs
Berlin: Ullstein Verlag, 2004

The Spontaneous Fulfillment of Desire
Harnessing the Infinite Power of Coincidence
New York: Random House Audio, 2003

Clarke, Ronald

Einstein: The Life and Times
New York: Avon Books, 1970

Craze, Richard

Feng Shui
Feng Shui Book & Card Pack
London: Thorsons, 1997

DeMeo, James

Heretic's Notebook
Emotions, Protocells, Ether-Drift and Cosmic Life Energy
with New Research Supporting Wilhelm Reich
Pulse of the Planet, #5 (2002)
Ashland, Oregon: Orgone Biophysical Research Laboratories, Inc., 2002

Nach Reich, Neue Forschungen zur Orgonomie
Sexualökonomie / Die Entdeckung der Orgonenergie
Herausgegeben zusammen mit Professor Bernd Senf, Berlin
Frankfurt/M: Zweitausendeins Verlag, 1997

Saharasia
The 4000 BCE Origins of Child Abuse, Sex-Repression, Warfare and Social Violence
in the Deserts of the Old World
Ashland, Oregon: Orgone Biophysical Research Laboratories, Inc., 1998

DiCarlo, Russell E. (Ed.)

Towards A New World View
Conversations at the Leading Edge
Erie, PA: Epic Publishing, 1996

Eden, Donna & Feinstein, David

Energy Medicine
New York: Tarcher/Putnam, 1998

The Energy Medicine Kit
Simple Effective Techniques to Help You Boost Your Vitality
Boulder, Co.: Sounds True Editions, 2004

The Promise of Energy Psychology
With David Feinstein and Gary Craig
Revolutionary Tools for Dramatic Personal Change
New York: Jeremy P. Tarcher/Penguin, 2005

Einstein, Albert

The World As I See It
New York: Citadel Press, 1993

Mein Weltbild
Berlin: Ullstein, 2005

Out of My Later Years
New York: Outlet, 1993

Ideas and Opinions
New York: Bonanza Books, 1988

Einstein sagt
Zitate, Einfälle, Gedanken
München: Piper, 2007

Albert Einstein Notebook
London: Dover Publications, 1989

Eisler, Riane

The Chalice and the Blade
Our history, Our future
San Francisco: Harper & Row, 1995

Kelch und Schwert, Unsere Geschichte, unsere Zukunft
Weibliches und männliches Prinzip in der Geschichte
Freiburg: Arbor Verlag, 2005

Sacred Pleasure: Sex, Myth and the Politics of the Body
New Paths to Power and Love
San Francisco: Harper & Row, 1996

The Partnership Way
New Tools for Living and Learning
With David Loye
Brandon, VT: Holistic Education Press, 1998

The Real Wealth of Nations
Creating a Caring Economics
San Francisco: Berrett-Koehler Publishers, 2008

Eliade, Mircea

Shamanism
Archaic Techniques of Ecstasy
New York: Pantheon Books, 1964

Emerson, Ralph Waldo

The Essays of Ralph Waldo Emerson
Cambridge, Mass.: Harvard University Press, 1987

Emoto, Masaru

The Hidden Messages in Water
New York: Atria Books, 2004

Die Botschaft des Wassers
Burgrain: Koha Verlag, 2008

The Secret Life of Water
New York: Atria Books, 2005

Die Heilkraft des Wassers
Burgrain: Koha Verlag, 2004

Fourcade, Jean-Michel

Analyse transactionnelle et bioénergie
Paris: Delarge, 1981

Franz Anton Mesmer

Franz Anton Mesmer und die Geschichte des Mesmerismus
Beiträge zum internationalen wissenschaftlichen Symposium
anlässlich des 250. Geburtstages von Mesmer
Stuttgart, 1985

Freud, Anna

War and Children
London: 1943

Freud, Sigmund

Three Essays on the Theory of Sexuality
in: The Standard Edition of the Complete Psychological Works of Sigmund Freud
London: Hogarth Press, 1953-54
Vol. 7, pp. 130 ff
(first published in 1905)

Drei Abhandlungen zur Sexualtheorie
Frankfurt/M: Fischer, 1991

The Interpretation of Dreams
New York: Avon, Reissue Edition, 1980
and in: The Standard Edition of the Complete Psychological Works of Sigmund Freud
(24 Volumes) ed. by James Strachey
New York: W. W. Norton & Company, 1976

Die Traumdeutung
Frankfurt/M: Fischer, 2005

Totem and Taboo
New York: Routledge, 1999
Originally published in 1913

Totem und Tabu
Einige Übereinstimmungen im Seelenleben der Wilden und der Neurotiker
Frankfurt/M: Fischer Verlag, 1972

Fromm, Erich

The Anatomy of Human Destructiveness
New York: Owl Book, 1992
Originally published in 1973

Anatomie der menschlichen Destruktivität
Berlin: Rowohlt, 1977

Escape from Freedom
New York: Owl Books, 1994
Originally published in 1941

Die Furcht vor der Freiheit
München: DTV Verlag, 1993

To Have or To Be
New York: Continuum International Publishing, 1996
Originally published in 1976

Haben oder Sein
Die seelischen Grundlagen einer neuen Gesellschaft
München: DTV Verlag, 2005

The Art of Loving
New York: HarperPerennial, 2000
Originally published in 1956

Die Kunst des Liebens
Berlin: Ullstein, 2005

Gerber, Richard

A Practical Guide to Vibrational Medicine
Energy Healing and Spiritual Transformation
New York: Harper & Collins, 2001

Gil, David G.

Societal Violence and Violence in Families
in: David G. Gil, Child Abuse and Violence
New York: Ams Press, 1928

Goethe, Johann Wolfgang von

The Theory of Colors
New York: MIT Press, 1970
First published in 1810

Goethes Farbenlehre
Leipzig: Seemann-Henschel Verlag, 1998

Goldenstein, Joyce

Einstein: Physicist and Genius
(Great Minds of Science)
New York: Enslow Publishers, 1995

Goldman, Jonathan & Goldman, Andi

Tantra of Sound
Frequencies of Healing
Charlottesville: Hampton Roads, 2005

Tantra des Klanges
Mehr Liebe und Intimität in der Partnerschaft
Mit CD
Hanau: Amra Verlag, 2009

Healing Sounds
The Power of Harmonies
Rochester: Healing Arts Press, 2002

Klangheilung
Die Schöpferkraft des Obertongesangs
Hanau: Amra Verlag, 2008

Healing Sounds
Principles of Sound Healing
DVD, 90 min.
Sacred Mysteries, 2004

Goleman, Daniel

Emotional Intelligence
New York, Bantam Books, 1995

EQ. Emotionale Intelligenz
München: DTV Verlag, 1997

Goswami, Amit

The Self-Aware Universe
How Consciousness Creates the Material World
New York: Tarcher/Putnam, 1995

Das Bewusste Universum
Wie Bewusstsein die materielle Welt erschafft
Stuttgart: Lüchow Verlag, 2007

Grant

Grant's Method of Anatomy
10th ed., by John V. Basmajian
Baltimore, London: Williams & Wilkins, 1980

Grof, Stanislav

Ancient Wisdom and Modern Science
New York: State University of New York Press, 1984

Beyond the Brain
Birth, Death and Transcendence in Psychotherapy
New York: State University of New York, 1985

LSD: Doorway to the Numinous
The Groundbreaking Psychedelic Research into Realms of the Human Unconscious
Rochester: Park Street Press, 2009

Published by Sirius-C Media Galaxy LLC, 2011

Psychologie transpersonnelle
Paris: Rocher, 1984

Realms of the Human Unconscious
Observations from LSD Research
New York: E.P. Dutton, 1976

The Cosmic Game
Explorations of the Frontiers of Human Consciousness
New York: State University of New York Press, 1998

The Holotropic Mind
The Three Levels of Human Consciousness
With Hal Zina Bennett
New York: HarperCollins, 1993

When the Impossible Happens
Adventures in Non-Ordinary Reality
Louisville, CO: Sounds True, 2005

Wir wissen mehr als unser Gehirn
Die Grenzen des Bewusstseins überschreiten
Freiburg: Herder, 2007

Hall, Manly P.

The Pineal Gland
The Eye of God
Article extracted from the book: Man the Grand Symbol of the Mysteries
Kessinger Publishing Reprint

The Secret Teachings of All Ages
Reader's Edition
New York: Tarcher/Penguin, 2003
Originally published in 1928

Hameroff, Newberg, Woolf, Bierman et al.

Consciousness
20 Scientists Interviewed
Director: Gregory Alsbury
5 DVD Box Set, 540 min.
New York: Alsbury Films, 2003

Harner, Michael

Ways of the Shaman
New York: Bantam, 1982
Originally published in 1980

Der Weg des Schamanen
Das praktische Grundlagenbuch zum Schamanismus
Genf: Ariston, 2007

Chamane
Les secrets d'un sorcier indien d'Amérique du Nord
Paris: Albin Michel, 1982

Hofmann, Albert

LSD, My Problem Child
Reflections on Sacred Drugs, Mysticism and Science
Santa Cruz, CA: Multidisciplinary Association for Psychedelic Studies, 2009
Originally published in 1980

LSD, Mein Sorgenkind
Die Entdeckung der 'Wunderdroge'
München: DTV Verlag, 1999

Holmes, Ernst

The Science of Mind
A Philosophy, A Faith, A Way of Life
New York: Jeremy P. Tarcher/Putnam, 1998
First Published in 1938

Houston, Jean

The Possible Human
A Course in Enhancing Your Physical, Mental, and Creative Abilities
New York: Jeremy P. Tarcher/Putnam, 1982

Hunt, Valerie

Infinite Mind
Science of the Human Vibrations of Consciousness
Malibu, CA: Malibu Publishing, 2000

Janov, Arthur

Primal Man
The New Consciousness
New York: Crowell, 1975

Das Neue Bewusstsein
Frankfurt/M: Fischer Verlag, 1988
Urausgabe 1975

Jung, Carl Gustav

Archetypen
München: DTV Verlag, 2001

Archetypes of the Collective Unconscious
in: The Basic Writings of C.G. Jung
New York: The Modern Library, 1959, 358-407

Collected Works
New York, 1959

Published by Sirius-C Media Galaxy LLC, 2011

Dialectique du moi et de l'inconscient
Paris, Gallimard, 1991

On the Nature of the Psyche
in: The Basic Writings of C.G. Jung
New York: The Modern Library, 1959, 47-133

Psychological Types
Collected Writings, Vol. 6
Princeton: Princeton University Press, 1971

Psychologie und Religion
München: DTV Verlag, 2001

Psychology and Religion
in: The Basic Writings of C.G. Jung
New York: The Modern Library, 1959, 582-655

Religious and Psychological Problems of Alchemy
in: The Basic Writings of C.G. Jung
New York: The Modern Library, 1959, 537-581

Symbol und Libido
Freiburg: Walter Verlag, 1987

Synchronizität, Akausalität und Okkultismus
Frankfurt/M: DTV, 2001

The Basic Writings of C.G. Jung
New York: The Modern Library, 1959

The Development of Personality
Collected Writings, Vol. 17
Princeton: Princeton University Press, 1954

The Meaning and Significance of Dreams
Boston: Sigo Press, 1991

The Myth of the Divine Child
in: Essays on A Science of Mythology
Princeton, N.J.: Princeton University Press Bollingen
Series XXII, 1969. (With Karl Kerenyi)

Traum und Traumdeutung
München: DTV Verlag, 2001

Two Essays on Analytical Psychology
Collected Writings, Vol. 7
Princeton: Princeton University Press, 1972
First published by Routledge & Kegan Paul, Ltd., 1953

Zur Psychologie westlicher und östlicher Religion
Fünfte Auflage
Olten: Walter Verlag, 1988

Karagulla, Shafica

The Chakras
Correlations between Medical Science and Clairvoyant Observation
With Dora van Gelder Kunz
Wheaton: Quest Books, 1989

Die Chakras und die feinstofflichen Körper des Menschen
Mit Dora van Gelder-Kunz
Grafing: Aquamarin Verlag, 1994

Kerner Justinus

F.A. Mesmer aus Schwaben
Frankfurt/M, 1856

Kiesewetter, Carl

Franz Anton Mesmer's Leben und Lehre
Leipzig, 1893

Kingston, Karen

Creating Sacred Space With Feng Shui
New York: Broadway Books, 1997

Krishnamurti, J.

Freedom From The Known
San Francisco: Harper & Row, 1969

The First and Last Freedom
San Francisco: Harper & Row, 1975

Education and the Significance of Life
London: Victor Gollancz, 1978

Commentaries on Living
First Series
London: Victor Gollancz, 1985

Commentaries on Living
Second Series
London: Victor Gollancz, 1986
Krishnamurti's Journal
London: Victor Gollancz, 1987

Krishnamurti's Notebook
London: Victor Gollancz, 1986

Published by Sirius-C Media Galaxy LLC, 2011

Beyond Violence
London: Victor Gollancz, 1985

Beginnings of Learning
New York: Penguin, 1986

The Penguin Krishnamurti Reader
New York: Penguin, 1987

On God
San Francisco: Harper & Row, 1992

On Fear
San Francisco: Harper & Row, 1995

The Essential Krishnamurti
San Francisco: Harper & Row, 1996

The Ending of Time
With Dr. David Bohm
San Francisco: Harper & Row, 1985

Kwok, Man-Ho

The Feng Shui Kit
London: Piatkus, 1995

Laing, Ronald David

Divided Self
New York: Viking Press, 1991

R.D. Laing and the Paths of Anti-Psychiatry
ed., by Z. Kotowicz
London: Routledge, 1997

The Politics of Experience
New York: Pantheon, 1983

Sagesse, déraison et folie
Paris: Seuil, 1986

Lakhovsky, Georges

La Science et le Bonheur
Longévité et Immortalité par les Vibrations
Paris: Gauthier-Villars, 1930

Le Secret de la Vie
Paris: Gauthier-Villars, 1929

Secret of Life
New York: Kessinger Publishing, 2003

L'étiologie du Cancer
Paris: Gauthier-Villars, 1929

L'Universion
Paris: Gauthier-Villars, 1927

Laszlo, Ervin

Holos. Die Welt der neuen Wissenschaften
Petersberg: Via Nova Verlag, 2002

Science and the Akashic Field
An Integral Theory of Everything
Rochester: Inner Traditions, 2004

Macroshift
Die Herausforderung
Frankfurt/M: Insel Verlag, 2003

Quantum Shift to the Global Brain
How the New Scientific Reality Can Change Us and Our World
Rochester: Inner Traditions, 2008

Science and the Reenchantment of the Cosmos
The Rise of the Integral Vision of Reality
Rochester: Inner Traditions, 2006

The Akashic Experience
Science and the Cosmic Memory Field
Rochester: Inner Traditions, 2009

The Chaos Point
The World at the Crossroads
Newburyport, MA: Hampton Roads Publishing, 2006

Leary, Timothy

Our Brain is God
Berkeley, CA: Ronin Publishing, 2001
Author Copyright 1988

Über die Kriminalisierung des Natürlichen
Löhrbach: Werner Pieper Verlag, 1990

Leboyer, Frederick

Birth Without Violence
New York, 1975

Pour une Naissance sans Violence
Paris: Seuil, 1974

Geburt ohne Gewalt
München: Kösel 1981

Published by Sirius-C Media Galaxy LLC, 2011

Cette Lumière d'où vient l'Enfant
Paris: Seuil, 1978

Inner Beauty, Inner Light
New York: Newmarket Press, 1997

Weg des Lichts
München: Kösel, 1991

Loving Hands
The Traditional Art of Baby Massage
New York: Newmarket Press, 1977

Sanfte Hände
Die Kunst der indischen Baby-Massage
München: Kösel, 1979

The Art of Breathing
New York: Newmarket Press, 1991

Liedloff, Jean

Continuum Concept
In Search of Happiness Lost
New York: Perseus Books, 1986
First published in 1977

Auf der Suche nach dem verlorenen Glück
Gegen die Zerstörung der Glücksfähigkeit in der frühen Kindheit
München: C.H. Beck Verlag, 2006

Lip, Evelyn

The Design & Feng Shui of Logos, Trademarks and Signboards
Singapore: Prentice Hall, 1995

Lipton, Bruce

The Biology of Belief
Unleashing the Power of Consciousness, Matter and Miracles
Santa Rosa, CA: Mountain of Love/Elite Books, 2005

Intelligente Zellen
Wie Erfahrungen unsere Gene steuern
Burgrain: Koha Verlag, 2006

Long, Max *Freedom*

The Secret Science at Work
The Huna Method as a Way of Life
Marina del Rey: De Vorss Publications, 1995
Originally published in 1953

Geheimes Wissen hinter Wundern
Die Entdeckung der HUNA-Lehre
Darmstadt: Schirner Verlag, 2006

Growing Into Light
A Personal Guide to Practicing the Huna Method,
Marina del Rey: De Vorss Publications, 1955

Lowen, Alexander

Angst vor dem Leben
Über den Ursprung seelischen Leides und den Weg zu einem reicheren Dasein
München: Goldmann Wilhelm, 1989

Bioenergetics
New York: Coward, McGoegham 1975

Bioenergetik
Therapie der Seele durch Arbeit mit dem Körper
Berlin: Rowohlt, 2008

Depression and the Body
The Biological Basis of Faith and Reality
New York: Penguin, 1992

Fear of Life
New York: Bioenergetic Press, 2003

Honoring the Body
The Autobiography of Alexander Lowen
New York: Bioenergetic Press, 2004

Joy
The Surrender to the Body and to Life
New York: Penguin, 1995

Liebe und Orgasmus
Persönlichkeitserfahrung durch sexuelle Erfüllung
München: Goldmann Wilhelm, 1993

Love and Orgasm
New York: Macmillan, 1965

Love, Sex and Your Heart
New York: Bioenergetics Press, 2004

Narcissism: Denial of the True Self
New York: Macmillan, Collier Books, 1983

Narzissmus
Die Verleugnung des wahren Selbst
München: Goldmann Wilhelm, 1992

Pleasure: A Creative Approach to Life
New York: Bioenergetics Press, 2004
First published in 1970

The Language of the Body
Physical Dynamics of Character Structure
New York: Bioenergetics Press, 2006
First published in 1958
New York: Sri Ramanasramam, 2002

Malinowski, Bronislaw

Crime und Custom in Savage Society
London: Kegan, 1926

Sex and Repression in Savage Society
London: Kegan, 1927

The Sexual Life of Savages in North West Melanesia
New York: Halycon House, 1929

Das Geschlechtsleben der Wilden in Nordwest-Melanesien
Liebe, Ehe und Familienleben bei den Eingeborenen der Trobriand Inseln,
Britisch-Neuguinea
Eschborn: Klotz Verlag, 2005

Mann, Edward W.

Orgone, Reich & Eros
Wilhelm Reich's Theory of Life Energy
New York: Simon & Schuster (Touchstone), 1973

McKenna, Terence

The Archaic Revival
San Francisco: Harper & Row, 1992

Food of The Gods
A Radical History of Plants, Drugs and Human Evolution
London: Rider, 1992

Die Speisen der Götter
Berlin: Synergia/Syntropia, 1996

The Invisible Landscape
Mind Hallucinogens and the I Ching
New York: HarperCollins, 1993
(With Dennis McKenna)

True Hallucinations
Being the Account of the Author's Extraordinary
Adventures in the Devil's Paradise
New York: Fine Communications, 1998

McTaggart, Lynne

The Field
The Quest for the Secret Force of the Universe
New York: Harper & Collins, 2002

Mead, Margaret

Sex and Temperament in Three Primitive Societies
New York, 1935

Meadows, Donella H.

Thinking in Systems
A Primer
White River, VT: Chelsea Green Publishing, 2008

Mehta, Rohit

J. Krishnamurti and the Nameless Experience
A Comprehensive Discussion of J. Krishnamurti's Approach to Life
Delhi: Motilal Banarsidass Publishers, 2002

Metzner, Ralph (Ed.)

Ayahuasca, Human Consciousness and the Spirits of Nature
ed. by Ralph Metzner, Ph.D
New York: Thunder's Mouth Press, 1999

The Psychedelic Experience
A Manual Based on the Tibetan Book of the Dead
With Timothy Leary and Richard Alpert
New York: Citadel, 1995

Miller, Alice

Four Your Own Good
Hidden Cruelty in Child-Rearing and the Roots of Violence
New York: Farrar, Straus & Giroux, 1983

Am Anfang war Erziehung
München: Suhrkamp Verlag, 2008
Erstmals publiziert im Jahre 1986

Pictures of a Childhood
New York: Farrar, Straus & Giroux, 1986

The Drama of the Gifted Child
In Search for the True Self
translated by Ruth Ward
New York: Basic Books, 1996

Das Drama des Begabten Kindes
Und die Suche nach dem wahren Selbst
München: Suhrkamp Verlag, 1983

Published by Sirius-C Media Galaxy LLC, 2011

Der gemiedene Schlüssel
München: Suhrkamp, 2007

Das verbannte Wissen
Frankfurt/M: Suhrkamp, 1988

Thou Shalt Not Be Aware
Society's Betrayal of the Child
New York: Noonday, 1998

Du Sollst Nicht Merken
Variationen über das Paradies-Thema
Neuauflage
München: Suhrkamp, 2005

The Political Consequences of Child Abuse
in: The Journal of Psychohistory 26, 2 (Fall 1998)

Montagu, Ashley

Touching
The Human Significance of the Skin
New York: Harper & Row, 1978

Körperkontakt
8. Auflage
Stuttgart: Klett/Cotta, 1995

Moore, Thomas

Care of the Soul
A Guide for Cultivating Depth and Sacredness in Everyday Life
New York: Harper & Collins, 1994

Die Seele Lieben
Tiefe und Spiritualität im täglichen Leben
München: Droemer Knaur, 1995

Murdock, G.

Social Structure
New York: Macmillan, 1960

Murphy, Joseph

The Power of Your Subconscious Mind
West Nyack, N.Y.: Parker, 1981, N.Y.: Bantam, 1982
Originally published in 1962

Die Macht Ihres Unterbewusstseins
München: Hugendubel, 2000

La puissance de votre subconscient
Genève: Ramón Keller, 1967

The Miracle of Mind Dynamics
New York: Prentice Hall, 1964

Miracle Power for Infinite Riches
West Nyack, N.Y.: Parker, 1972

The Amazing Laws of Cosmic Mind Power
West Nyack, N.Y.: Parker, 1973

Secrets of the I Ching
West Nyack, N.Y.: Parker, 1970

Think Yourself Rich
Use the Power of Your Subconscious Mind to Find True Wealth
Revised by Ian D. McMahan, Ph.D.
Paramus, NJ: Reward Books, 2001

Das Erfolgsbuch
Wie sie alles im Leben erreichen können
Hamburg: Heyne Verlag, 2002

Wahrheiten die ihr Leben verändern
Dr. Joseph Murphys Vermächtnis
München: Hugendubel, 1996

Murphy, Michael
The Future of the Body
Explorations into the Further Evolution of Human Nature
New York: Jeremy P. Tarcher/Putnam, 1992

Der Quanten-Mensch
München: Ludwig Verlag, 1996

Narby, Jeremy
The Cosmic Serpent
DNA and the Origins of Knowledge
New York: J. P. Tarcher, 1999

Die Kosmische Schlange
Auf den Pfaden der Schamanen zu den Ursprüngen modernen Wissens
Stuttgart: Klett-Cotta, 2007

Nau, Erika
Self-Awareness Through Huna
Virginia Beach: Donning, 1981

Selbstbewusst durch Huna
Die magische Weisheit Hawaiis
2. Auflage
Basel: Sphinx Verlag, 1989

Published by Sirius-C Media Galaxy LLC, 2011

Newton, Michael

Life Between Lives
Hypnotherapy for Spiritual Regression
Woodbury, Minn.: Llewellyn Publications, 2006

Odent, Michel

Birth Reborn
What Childbirth Should Be
London: Souvenir Press, 1994

The Scientification of Love
London: Free Association Books, 1999

Die Wurzeln der Liebe
Wie unsere wichtigsten Emotionen entstehen
Olten: Walter Verlag, 2001

Primal Health
Understanding the Critical Period Between Conception and the First Birthday
London: Clairview Books, 2002
First Published in 1986 with Century Hutchinson in London

La Santé Primale
Paris: Payot, 1986

Die sanfte Geburt
Die Leboyer-Methode in der Praxis
Bergisch-Gladbach: Lübbe Verlag, 2001

The Functions of the Orgasms
The Highway to Transcendence
London: Pinter & Martin, 2009

Ollendorf-Reich, Ilse

Wilhelm Reich, A Personal Biography
New York, St. Martins Press, 1969

Wilhelm Reich
Vorwort von A.S. Neill
München, Kindler, 1975

Ong, Hean-Tatt

Amazing Scientific Basis of Feng-Shui
Kuala Lumpur: Eastern Dragon Press, 1997

Ostrander, Sheila & Schroeder, Lynn

Superlearning 2000
New York: Delacorte Press, 1994

Superlearning
Die revolutionäre Lernmethode
München: Scherz Verlag, 1979

Supermemory
New York: Carroll & Graf, 1991

SuperMemory
Der Weg zum optimalen Gedächtnis
München: Goldmann, 1996

Pert, Candace B.

Molecules of Emotion
The Science Behind Mind-Body Medicine
New York: Scribner, 2003

Ponder, Catherine

The Healing Secrets of the Ages
Marine del Rey: DeVorss, 1985

Porteous, Hedy S.

Sex and Identity
Your Child's Sexuality
Indianapolis: Bobbs-Merrill, 1972

Prescott, James W.

Affectional Bonding for the Prevention of Violent Behaviors
Neurobiological, Psychological and Religious/Spiritual Determinants
in: Hertzberg, L.J., Ostrum, G.F. and Field, J.R., (Eds.)
Violent Behavior
Vol. 1, Assessment & Intervention, Chapter Six
New York: PMA Publishing, 1990

Alienation of Affection
Psychology Today, December 1979

Body Pleasure and the Origins of Violence
Bulletin of the Atomic Scientists, 10-20 (1975)

Deprivation of Physical Affection as a Primary Process in the
Development of Physical Violence A Comparative and Cross-Cultural Perspective,
in: David G. Gil, ed., Child Abuse and Violence
New York: Ams Press, 1979

Early somatosensory deprivation as an ontogenetic process in the abnormal
development of the brain and behavior,
in: Medical Primatology, ed. by I.E. Goldsmith and J. Moor-Jankowski,
New York: S. Karger, 1971

Genital Mutilation of Children
Failure of Humanity and Humanism

Published by Sirius-C Media Galaxy LLC, 2011

Unprinted Essay (2005)
http://www.violence.de/prescott/letters/CIRC_CONGRESS_MONTAGUE_9.30.05.html

Genital Pain vs. Genital Pleasure
Why the One and not the Other
The Truth Seeker, July/August 1989, pp. 14-21
http://www.violence.de/prescott/truthseeker/genpl.html

How Culture Shapes the Developing Brain and the Future of Humanity
A Brief Summary of the research which links brain abnormalities and violence
to an absence of nurturing and bonding very early in childhood,
in: Touch the Future: Optimum Learning Relationships for Children & Adults
Spring 2002 (Ed. by Michael Mendizza)
Nevada City, CA, 2002

Invited Commentary: Central nervous system functioning in altered sensory environments,
in: M.H. Appley and R. Trumbull (Eds.), *Psychological Stress,*
New York: Appleton-Century Crofts, 1967

Our Two Cultural Brains: Neurointegrative and Neurodissociative
http://www.violence.de/prescott/letters/Our_Two_Cultural_Brains.pdf

Phylogenetic and ontogenetic aspects of human affectional development,
in: Progress in Sexology, Proceedings of the 1976 International Congress of Sexology,
ed. by R. Gemme & C.C. Wheeler
New York: Plenum Press, 1977

Prevention or Therapy and the Politics of Trust
Inspiring a New Human Agenda
in: *Psychotherapy and Politics International*
Volume 3(3), pp. 194-211
London: John Wiley, 2005

Sex and the Brain
Midcontinent & Eastern Regions, June 13-16, 2002
Big Rapids, MI: Society for Cross-Cultural Research, 32nd Annual Meeting, 2005
http://www.violence.de/archive.shtml

Sixteen Principles for Personal, Family and Global Peace
The Truth Seeker, March/April 1989
http://www.violence.de/prescott/letters/Sixteen_Principles.pdf

Somatosensory affectional deprivation (SAD) theory of drug and alcohol use,
in: Theories on Drug Abuse: Selected Contemporary Perspectives,
ed. by Dan J. Lettieri, Mollie Sayers and Helen Wallenstien Pearson,
NIDA Research Monograph 30, March 1980
Rockville, MD: National Institute on Drug Abuse, Department of Health and Human
Services, 1980

The Origins of Human Love and Violence
Pre- and Perinatal Psychology Journal
Volume 10, Number 3:
Spring 1996, pp. 143-188The Origins of Love and Violence
Sensory Deprivation and the Developing Brain
Research and Prevention (DVD)

http://violence.de
http://ttfuture.org/violence
http://montagunocircpetition.org

Radin, Dean
The Conscious Universe
The Scientific Truth of Psychic Phenomena
San Francisco: Harper & Row, 1997

Entangled Minds
Extrasensory Experiences in a Quantum Reality
New York: Paraview Pocket Books, 2006

Raknes, Ola
Wilhelm Reich and Orgonomy
Oslo: Universitetsforlaget, 1970

Wilhelm Reich und die Orgonomie
Eine Einführung in die Wissenschaft von der Lebensenergie
Frankfurt/M: Nexus, 1983

Randall, Neville
Life After Death
London: Robert Hale, 1999

Rausky, Franklin
Mesmer ou la révolution thérapeutique
Paris, 1977

Redfield, James
The Tenth Insight
Holding the Vision
New York: Warner Books, 1996

The Celestine Prophecy
New York: Warner Books, 1995

Die Vision von Celestine
Berlin: Ullstein, 2004

Reich, Wilhelm

A Review of the Theories, dating from The 17th Century,
on the Origin of Organic Life
by Arthur Hahn, Literature Assistant at the Institut für Sexualökonomische
Lebensforschung, Biologisches Laboratorium, Oslo, 1938
©1979 by Mary Boyd Higgins as Director of the Wilhelm Reich Infant Trust
XEROX Copy from the Wilhelm Reich Museum

Children of the Future
On the Prevention of Sexual Pathology
New York: Farrar, Straus & Giroux, 1983
First published in 1950

CORE (Cosmic Orgone Engineering)
Part I, Space Ships, DOR and DROUGHT
©1984, Orgone Institute Press
XEROX Copy from the Wilhelm Reich Museum

Der Einbruch der sexuellen Zwangsmoral
Frankfurt/M: Fischer, 1981

Die Entdeckung des Orgons II
Der Krebs
Frankfurt/M: Fischer, 1981
Köln: Kiepenheuer & Witsch, 1984

Die Funktion des Orgasmus
Sexualökonomische Grundprobleme der biologischen Energie
Köln: Kiepenheuer & Witsch, 1987

Die Massenpsychologie des Faschismus
Frankfurt/M: Fischer, 1974

Die sexuelle Revolution
Frankfurt/M: Fischer, 1966

Early Writings 1
New York: Farrar, Straus & Giroux, 1975

Ether, God & Devil & Cosmic Superimposition
New York: Farrar, Straus & Giroux, 1972
Originally published in 1949

Frühe Schriften 1
Aus den Jahren 1920-1925
Frankfurt/M: Fischer, 1983

Frühe Schriften 2
Genitalität in der Theorie und Therapie der Neurose
Frankfurt/M: Fischer, 1985

Genitality in the Theory and Therapy of Neurosis
©1980 by Mary Boyd Higgins as Director of the Wilhelm Reich Infant Trust

Leidenschaften der Jugend
Köln: Kiepenheuer & Witsch, 1984

L'irruption de la morale sexuelle
Paris: Payot, 1972

Menschen im Staat
Frankfurt/M: Nexus, 1982

People in Trouble
©1974 by Mary Boyd Higgins as Director of the Wilhelm Reich Infant Trust

Record of a Friendship
The Correspondence of Wilhelm Reich and A. S. Neill
New York, Farrar, Straus & Giroux, 1981

Selected Writings
An Introduction to Orgonomy
New York: Farrar, Straus & Giroux, 1973

The Bioelectrical Investigation of Sexuality and Anxiety
New York: Farrar, Straus & Giroux, 1983
Originally published in 1935

The Bion Experiments
reprinted in *Selected Writings*
New York: Farrar, Straus & Giroux, 1973

The Cancer Biopathy (The Orgone, Vol. 2)
New York: Farrar, Straus & Giroux, 1973

The Function of the Orgasm (The Orgone, Vol. 1)
Orgone Institute Press, New York, 1942

The Invasion of Compulsory Sex Morality
New York: Farrar, Straus & Giroux, 1971
Originally published in 1932

The Leukemia Problem: Approach
©1951, Orgone Institute Press
Copyright Renewed 1979
XEROX Copy from the Wilhelm Reich Museum

The Mass Psychology of Fascism
New York: Farrar, Straus & Giroux, 1970
Originally published in 1933

The Orgone Energy Accumulator
Its Scientific and Medical Use
©1951, 1979, Orgone Institute Press
XEROX Copy from the Wilhelm Reich Museum

Published by Sirius-C Media Galaxy LLC, 2011

The Schizophrenic Split
©1945, 1949, 1972 by Mary Boyd Higgins as Director of the
Wilhelm Reich Infant Trust
XEROX Copy from the Wilhelm Reich Museum

The Sexual Revolution
©1945, 1962 by Mary Boyd Higgins as Director of the Wilhelm Reich Infant Trust

Zeugnisse einer Freundschaft
Der Briefwechsel zwischen Wilhelm Reich und A.S.
Neill (1936-1957)
Köln: Kiepenheuer & Witsch, 1986

Richet, Charles

Metapsychical Phenomena
Methods and Observations
Kessinger Publishing Reprint Edition, 2004
Originally published in 1905

Roberts, Jane

The Nature of Personal Reality
New York: Amber-Allen Publishing, 1994
First published in 1974

Die Natur der Persönlichen Realität
Ein neues Bewusstsein als Quelle der Kreativität
München: Kailash Verlag, 2007

The Nature of the Psyche
Its Human Expression
New York, Amber-Allen Publishing, 1996
First published in 1979

Die Natur der Psyche
Ihr menschlicher Ausdruck in Kreativität, Liebe, Sexualität
Genf: Ariston Verlag, 1985

Die Natur der Psyche
Ihr menschlicher Ausdruck in Kreativität, Liebe, Sexualität
München: Kailash Verlag, 2008

Roman, Sanaya

Opening to Channel
How To Connect With Your Guide
New York: H.J. Kramer, 1987

Zum Höheren Selbst Erwachen
Das Herz dem Bewusstsein des Lichts öffnen
Genf: Ansata Verlag, 2003

Ruiz, Don Miguel

The Four Agreements
A Practical Guide to Personal Freedom
San Rafael, CA: Amber Allen Publishing, 1997

The Mastery of Love
A Practical Guide to the Art of Relationship
San Rafael, CA: Amber Allen Publishing, 1999

The Voice of Knowledge
A Practical Guide to Inner Peace
With Janet Mills
San Rafael, CA: Amber Allen Publishing, 2004

SantoPietro, Nancy

Feng Shui, Harmony by Design
How to Create a Beautiful and Harmonious Home,
New York: Putnam-Berkeley, 1996

Schlipp, Paul A. (Ed.)

Albert Einstein
Philosopher-Scientist
New York: Open Court Publishing, 1988

Schrenck-Notzing, Albert von

Phenomena of Materialization
A Contribution to the Investigation of Mediumistic Teleplastics
Perspectives in Psychical Research
New York: Kegan Paul, 1920

Schultes, Richard Evans, et al.

Plants of the Gods
Their Sacred, Healing, and Hallucinogenic Powers
New York: Healing Arts Press
2nd edition, 2002

Die Pflanzen der Götter
Die magischen Kräfte der Rausch- und Giftgewächse
München: AT Verlag, 1998

Senf, Bernd

Die Wiederentdeckung des Lebendigen
Aachen: Omega, 2003
Erstmals veröffentlicht 1996 mit Zweitausendeins Verlag in Frankfurt/M

Nach Reich: Neue Forschungen zur Orgonenergie
Sexualökonomie / Die Entdeckung der Orgonenergie
Herausgegeben zusammen mit Professor James DeMeo, Ashland, Oregon, USA
Frankfurt/M: Zweitausendeins Verlag, 1997

Published by Sirius-C Media Galaxy LLC, 2011

Sepper, Dennis L.

Goethe Contra Newton
Polemics and the Project of a New Science of Color
Cambridge: Cambridge University Press, 1988

Sharaf, Myron

Fury on Earth
A Biography of Wilhelm Reich
London: André Deutsch, 1983

Wilhelm Reich
Der heilige Zorn des Lebendigen
Berlin: Simon & Leutner, 1994

Sheldrake, Rupert

A New Science of Life
The Hypothesis of Morphic Resonance
Rochester: Park Street Press, 1995

Das Schöpferische Universum
Die Theorie des morphogenetischen Feldes
Neue und erweiterte Auflage
Berlin: Ullstein, 2009

Simonton, O. Carl et al.

Getting Well Again
Los Angeles: Tarcher, 1978

Strassman, Rick

DMT: The Spirit Molecule
A doctor's revolutionary research into the biology of near-death
and mystical experiences
Rochester: Park Street Press, 2001

Talbot, Michael

The Holographic Universe
New York: HarperCollins, 1992

Das holographische Universum
Die Welt in neuer Dimension
München: Droemer Knaur, 1994

Tansley, David V.

Chakras, Rays and Radionics
London: Daniel Company Ltd., 1984

Targ, Russell & Katra, Jane

Miracles of Mind
Exploring Nonlocal Consciousness and Spiritual Healing
Novato, CA: New World Library, 1999

Tarnas, Richard

Cosmos and Psyche
Intimations of a New World View
New York: Plume, 2007

The Passion of the Western Mind
Understanding the Ideas that have Shaped Our World View
New York: Ballantine Books, 1993

Tart, Charles T.

Altered States of Consciousness
A Book of Readings
Hoboken, N.J.: Wiley & Sons, 1969

Tatar, Maria M.

Spellbound: Studies on Mesmerism and Literature
Princeton, N.Y., 1978

Textor, R. B.

A Cross-Cultural Summary
New Haven, Human Relations Area Files (HRAF)
Press, 1967

Tiller, William A.

Conscious Acts of Creation
The Emergence of a New Physics
Associated Producers, 2004 (DVD)

Psychoenergetic Science
New York: Pavior, 2007

Conscious Acts of Creation
New York: Pavior, 2001

Tischner, Rudolf

F.A. Mesmer
München, 1928

Todaro-Franceschi, Vidette

The Enigma of Energy
Where Science and Religion Converge
New York: Crossroad Publishing, 1991

Published by Sirius-C Media Galaxy LLC, 2011

Too, Lillian

Feng Shui
Kuala Lumpur: Konsep Books, 1994

What the Bleep Do We Know!?

See Arntz, William

Wilber, Ken

Sex, Ecology, Spirituality
The Spirit of Evolution
Boston: Shambhala, 2000

Quantum Questions
Mystical Writings of The World's Greatest Physicists
Boston: Shambhala, 2001

Wolf, Fred Alan

Taking the Quantum Leap
The New Physics for Nonscientists
New York: Harper & Row, 1989

Der Quantensprung ist keine Hexerei
Frankfurt/M: Fischer Verlag, 1990

Parallel Universes
New York: Simon & Schuster, 1990

The Dreaming Universe
A Mind-Expanding Journey into the Realm Where Psyche and Physics Meet
New York: Touchstone, 1995

The Eagle's Quest
A Physicist Finds the Scientific Truth At the Heart of the Shamanic World
New York: Touchstone, 1997

Die Physik der Träume
Frankfurt/M: DTV Verlag, 1997

Mind into Matter
A New Alchemy of Science and Spirit
New York: Moment Point Press, 2000

Wydra, Nancilee

Feng Shui
The Book of Cures
Lincolnwood: Contemporary Books, 1996

Zukav, Gary

The Dancing Wu Li Masters
An Overview of the New Physics
New York: HarperOne, 2001

Die tanzenden Wu Li Meister
Der östliche Pfad zum Verständnis der modernen Physik
Vom Quantensprung zum schwarzen Loch
Berlin: Rowohlt, 2000

Published by Sirius-C Media Galaxy LLC, 2011

FROM THE SAME AUTHOR

A Bibliography

You can search publications from here:
http://ipublica.com/books/

For audio books and music, you can start here:
http://ipublica.com/audio/

All paperbacks, audio downloads, audio book compact discs, music downloads and music compact discs as well as Kindle books are referenced on the site.

For free podcasts search iTunes under my author name.

For quoting any of my publications, please use the following form:
Pierre F. Walter, [Title]: [Subtitle], Newark: Sirius–C Media Galaxy LLC, 2011

Web Presence

Pierre F. Walter on the Web

Sites

http://authoryourlife.com

http://ipublica.com

http://ipublica.net

http://ipublica.org

http://ipublica.tv

Video Channels

http://youtube.com/user/ipublica

http://youtube.com/user/authoryourlife

http://vimeo.com/pierrefwalter/channels

http://ipublica.blip.tv/

http://authoryourlife.blip.tv/

http://emosexuality.blip.tv/

http://pierrefwalter.blip.tv/

Published by Sirius-C Media Galaxy LLC, 2011

www.ingramcontent.com/pod-product-compliance
Lightning Source LLC
Chambersburg PA
CBHW081432170526
45166CB00008B/2175